Better Angling with Simple Science

Better Angling with Simple Science
by Mary M Pratt

Fishing News (Books) Ltd
23 Rosemount Avenue, West Byfleet
Surrey, England

ISBN 085238 069 0

Printed in Great Britain by
The Whitefriars Press Ltd, London and Tonbridge

Contents

	Page
List of Figures	vii
Preface	ix
Introduction	xi

Chapter 1 – Living Things — 1

What is life and what is a living organism? — 1

Comparison of a living organism with an internal combustion engine. — 1

Food – its manufacture and use. — 4

The cellular structure of living things. — 5

Different kinds of living things and ways of naming them. — 6

The fish as a living organism – its respiration, feeding, movement, excretion, reproduction and growth. — 12

Chapter 2 – Lakes and Rivers — 22

The water – and do fish drink it? – and an experiment to show why freshwater fish do not drink it. — 22

Water density and support. — 25

Oxygen – summer and winter kills and the conditions producing deoxygenation. — 25

Dissolved substances and pH – the importance of pH in freshwater ecology and methods of measuring it. — 28

Temperature and lake stratification – fishing into the thick end of the wedge. — 31

Light — 35

Physical characteristics of lakes and rivers. — 35

Chapter 3 – Food — 39

Food webs. — 39

Gut analysis and the scientific study of feeding habits. — 42

Practical issues in the management of fishing waters.   46
Baits – artificial and natural and their presentation.   47
Energy flow and ideas about ecosystems and the efficiency   49
    of natural systems.
Ideas about productivity.   52

Chapter 4 – The Lives of Fishes   53
The activity, rates of living and adaptations of different   53
    species to different conditions with special reference
    to temperature and oxygen.
Shapes of fish and their swimming activities.   58
Breeding habits.   61

Chapter 5 – Fish Senses and Behaviour   63
The Senses and how they work – sight, lateral line sense,   64
    hearing, olfactory sense, *etc.*
Ways of studying fish behaviour.   73
Different kinds of behaviour – inherited and learned.   74
Can fish learn to avoid being caught?   77
Can fish communicate?   79

Chapter 6 – Growth, Population Numbers and   82
Productivity
Methods of determining age and patterns of growth – scale   83
    reading.
Methods of estimating population numbers in fishery   86
    surveys.
The study of productivity and its practical applications,   88
    eg in dealing with problems of stunted fish, stocking
    and manipulation of numbers *etc.*

Chapter 7 – Maintaining Healthy Angling Waters   97
Fish disease – what is disease and what are its causes?   98
Some common parasitic diseases and methods of   101
    controlling them.
Pollution – different kinds of pollution.   105
How can anglers detect pollution and what is being done   108
    about it?
Water abstraction and the problems it creates.   114
Conservation – the angler and the countryside.   115

List of Figures

		Page
Fig 1	Structure of a protein molecule.	3
Fig 2	Features of an animal cell under microscope.	5
Fig 3	Evolutionary tree of plant and animal kingdom.	10, 11
Fig 4	External features of a typical teleost fish.	12
Fig 5	Internal organization and structure of a typical teleost fish.	13
Fig 6	Arrangement of blood vessels and blood circulation in the gill region.	14
Fig 7	Structure and functioning of the gills in a typical teleost fish.	15
Fig 8	Diagram illustrating the potato experiment.	23
Fig 9	Temperature stratification in a fairly deep lake.	32
Fig 10	Simplified food web in lake, river pool or stream.	40
Fig 11	Illustrating the food web operating in a productive pool.	42, 43
Fig 12	Showing how the food of the roach changes with age.	45
Fig 13	Illustrating the energy flow of a freshwater ecosystem.	49
Fig 14	Tolerance of temperature in various fishes.	56
Fig 15	Showing the 'compromise' torpedo fish.	59
Fig 16	Outlines and sections of various fishes illustrating different habits of life.	60
Fig 17	Vertical section through a typical teleost fish eye.	66
Fig 18	What a fish sees looking up when the water surface is perfectly calm.	67
Fig 19	Structure of the lateral line system.	68
Fig 20	Structure of the ear of a cyprinid fish.	69
Fig 21	Photographic impression of a bream scale.	84
Fig 22	Life cycle of the eye fluke.	99
Fig 23	Life cycle of the fish tapeworm.	100
Plates 1	Organisms useful in indicating the presence and severity of organic pollution.	109

Preface

This book has emerged from a series of classes in freshwater biology run by the Workers Educational Association for anglers in the North Staffordshire area. It was the interest and enthusiasm shown by the members of my classes, and the questions they have asked, which stimulated me to write something down in book form.

My first acknowledgement is then to my angler friends in Staffordshire from whom I learnt so much. The 'wise angler and naturalist' mentioned on page 79 is Richard Crudgington to whom I am greatly indebted for allowing me to quote from his angling articles and for help and advice in the preparation of the book.

I am also most grateful to Mrs M E Varley for reading a large part of the manuscript and for very helpful advice on scientific matters. Thanks are due also to Miss R M Badcock and to Dr P Chevins who have helped with parts of the text and to Dr T Macan, Mr J Clegg, Mr T Bagenal and Dr G Fearnley for valuable discussions and correspondence about fish, fishermen and fishing. I should like also to thank Mr Gerald Burgess for taking the photographs for Figure 21 and Plate 1. Thanks are also due to the University of Keele Biology Department for the use of these photographs.

Finally I must not fail to thank my family for their encouragement and patience with me while this book was being written and in particular my husband for invaluable assistance in reading the manuscript as a non-scientific guinea-pig.

Introduction

Not long ago I read of an angler who had made a curious discovery. He had been puzzled for some time by the fact that certain pegs on a particular stretch of water consistently provided good match results. So he set out on a piece of detective work. He laid aside rod and keep-net and took up the use of dragline and scoop. With these he scoured the bottom and carefully examined what he hauled up. On analysing his catches he found that in the spots which fished best were shrimps and blood worms in enormous numbers. What is more, the patches of these creatures corresponded in turn with patches of fine blanket weed. Putting two and two together it looked to him as though blanket weed indicated the presence of fish – the fish no doubt having been attracted there to feed on the shrimps and bloodworms. Continuing his search he began to find patches of weed in unexpected places, and, sure enough, the fish were there too. Here was an angler turned scientist who found that a brief excursion into freshwater ecology was greatly to his advantage.

Now here I must say that I am sure angling is first and foremost an art. The scientist must tread warily when venturing into the angler's preserve for there is much about the successful fisherman which defies scientific analysis. But, just as the motor-racing ace and the mechanical engineer have a working partnership or just as the painter and the expert in optics or visual perception have a good deal of interest to say to each other, so the angler and the biologist should find discussion fruitful. And might I throw in two ingredients for a good discussion – the scientist must beware of falling into his sometimes unhappy habit of wrapping things up in too many layers of technicalities, and the angler must be prepared to have a sharp eye cast on some of his tales!

Biology is the study of living things and ecology (a branch of biology) is the scientific study of living things in relation to their environment. The biologist works, as do all scientists, by observing and recording facts, putting forward an idea (or hypothesis) to explain the observations, and then testing the hypothesis by making further recordings and carrying out experiments. If the experiments support the hypothesis it may become established as a theory or law.

This is just the sort of thing that the angler I described above was doing and I am sure that this way one can shed more light on things than by relying on folk-lore or guess work. This is not to say that folk-lore and guess work are always wrong – only that they are less reliable for the purposes of explaining and predicting natural events.

Scientific activity has always had by-products beneficial to man. For example we can now keep ourselves warm with central heating, wet-suits and nylon fur; we can expect to live longer and find for ourselves a greater variety of hobbies and interests. Of course science has sometimes made a mess of things. Technology in the hands of greedy men has certainly perpetrated gross mistakes and exposed us to dangers. But it is equally true to say that science can in turn help to put things right. In doing this the science of ecology has a particularly important part to play. By no means less important is the sheer thrill and satisfaction of the pursuit of science for its own sake; those who embark on a path of scientific enquiry very often do so for the personal pleasures and rewards it brings.

There are many aspects of the sciences of fish biology and freshwater ecology which might be of direct interest to anglers. An insight into the operation of natural processes in freshwaters and an understanding of the ways in which living things interact with each other and with the environment, should help members of angling clubs to manage their waters and to provide good sport for the increasing numbers of people who are demanding it. To promote healthy fish populations, to increase the numbers of good sized fish, to recognise signs of pollution and to deal with it, are all part of a club's activities, and in all these purposes scientific approaches can help. Not least, for the ordinary angler, an interest in the private lives of the fish he is after should help to make the many hours of waiting by the water's edge more enjoyable.

On the other side of the coin the angler can assist the scientist. Part of the angler's art is to be constantly watchful and observant. On the spot, and ideally placed to notice interesting aspects of the habits of fishes, changes in water conditions, effects of the passing seasons or changing weather, he can be a useful ally to the biologist in his laboratory. The observations of anglers could well be useful to ecologists in suggesting lines of further investigation and a challenge in providing practical problems with which to grapple.

This book is not an instant guide to filling the bag with record specimens. I do hope, however, that it gives fairly straightforward explanations of some of the things which biologists have to say about fish and their ways which are likely to be of interest to fishermen and which may be of direct use in managing and maintaining good angling waters. Above all I hope that it will contribute in some small way to the hours of enjoyment with rod and line.

Chapter 1 – Living Things

WHAT IS LIFE?

As any angler is well aware, his quarry is very much alive. Every fisherman must have a tale to tell about the fight put up on some occasion or other, and most will have experienced the tremendous strength of a big fish's muscular power. Perhaps many is the still summer evening you have watched a shoal going about the business of feeding or apparently idling aimlessly to and fro. How do these creatures tick?

Human beings have been puzzling over the nature of life ever since they began to think about themselves and their surroundings. The descriptions and explanations which follow give a glimpse of some of the ideas which biologists now have about living organisms. We cannot possibly in this book delve into the details of all the incredible complexities life presents to us. Fortunately, however, over the years of painstaking research, biologists have found that all living things have a good deal in common. It is thus possible to create a picture of the basic functions of a plant or animal and to make some generalisations about the way in which they keep going.

THE LIVING ORGANISM AS A MACHINE

Before you are put off by the rather forbidding heading to this section let me assure you that thinking about living creatures as machines is not necessarily the whole story – it is simply a rather useful sort of comparison to make when trying to understand how living things work. When you are asked 'How on earth does that work?' very often the easiest thing to do is to try and compare it with something else which is familiar. Indeed, the scientist when tackling a new problem might well say to himself 'Do I know anything at all like this which might shed some light on the question?'

It doesn't take long to observe that a living creature needs constant supplies of oxygen and fuel (in the case of ourselves the

1

latter needs supplying at extraordinarily frequent intervals – or so it seems to me when I am supplying the family!). 'How like my car' you might then say – especially if it is an old banger which only does fifteen miles to the gallon! Indeed, comparing the living organism with the internal combustion engine turns out to be very illuminating.

An internal combustion engine is constructed out of matter (materials such as metal and plastic) and is designed to release energy from fuel (matter in which large quantities of energy are stored). The stored energy in the fuel is released and diverted into doing the work of turning a crankshaft and this happens by a process of combustion – simply the burning up of the fuel in the presence of oxygen.

For the moment then, we are going to think of our fish as a machine specially designed for releasing energy from fuel (food) and capable of using that energy to do the work of moving around, being aware of its surroundings, repairing itself, getting rid of waste, growing and reproducing its own kind. An oxygen supply and a fuel supply are essential for its operation just as they are for the internal combustion engine.

The construction of the living machine is immensely complicated. The building blocks are organic chemical substances the most important of which are the proteins. To illustrate the complexity of living materials in comparison with the stuff of which a car is made we can compare one molecule of iron with part of one molecule of a simple protein (see Figure 1). (Molecule is the chemist's term for the smallest structural unit of a substance capable of existing freely and unbound to any other particle. All matter is thought of by chemists as consisting of billions of these tiny particles massed together.) A molecule of iron consists simply of a single atom on its own. A protein molecule consists of many amino acid units linked together, each amino acid being itself a complex group of atoms of carbon, nitrogen, hydrogen and oxygen. The amino acid units are linked, as shown in the diagram, to form a chain and this chain may then be twisted and folded to form a more compact structure. The structural units of living material are thus much larger and a great deal more complex than those of which many man-made things are constructed.

The most important fuel from which living organisms are capable of releasing energy is glucose, again an organic substance with large complex molecules, although glucose itself

2

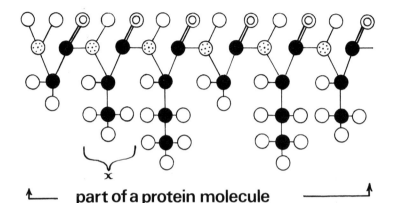

iron molecule

= atom of iron

part of a protein molecule

●= carbon atom ⊙= nitrogen atom ◎= oxygen atom

○= hydrogen atom

Fig 1 Illustrates the complex structure of a protein molecule in comparison with the simplicity of a molecule of iron. Shown in the diagram is only a small part of a protein molecule. It is in the form of a chain of sub-units (x), each sub-unit being an amino acid. A complete chain, which may be many times the length shown in the diagram, is usually twisted and folded to form a protein molecule. Such a molecule is one of the structural units of which living material is made, just as iron is one of the structural units of which steel is made. Clearly, living material is very much more complex than something like steel.

does not have such a complex make-up as a protein – it is more akin to a single amino acid in the size of its molecules.

The energy releasing process, the slow combustion of glucose in the presence of oxygen, is called respiration. Respiration can be represented as a chemical equation as follows

glucose + oxygen = carbon dioxide + water + energy

$$C_6H_{12}O_6 + 6O_2 = 6CO_2 + 6H_2O + energy$$

These principles apply to any living organism one might care to name; the cat on your lap, the fish in your pool, the sycamore tree in the park, the chrysanthemum in your garden, the germs in your throat all have the properties of a machine which converts stored chemical energy into energy available for doing work. The tree and the flowering plant may not appear to be exactly working themselves to the bone, but, nevertheless, they are earning a modest living busily transporting water from roots to leaves, producing foliage and flowers in their yearly spurt of furious growth, *etc*. It is in many ways an insult to compare them with that human invention the internal combustion engine, but it is nevertheless a helpful and illuminating comparison to make when one is trying to understand life.

THE MANUFACTURE OF COMPONENT PARTS AND FUEL

Having gone a few steps on the way to solving the mystery of the living organism, a very important question now arises. Where do the constructional materials, the protein molecules, come from, and how does the organism refuel? The answer is that food is the source of these requisites. Now it is obvious how the cat, the fish, even possibly the germ, get their foods, but it is perhaps not so obvious how the plants get hold of their glucose and proteins. In fact they manufacture their own food. For the green plant there is no hanging around waiting for bricks and mortar, petrol or oil to be delivered! They make for themselves all the building materials and fuel that they need and all that they require is a supply of carbon dioxide, water and certain simple nutrient salts from the soil. They do this by a chemical process called photosynthesis which involves the combining of carbon dioxide and water, in the presence of sunlight and a green pigment, chlorophyll, to make glucose. Again this can be represented by a chemical equation:

$$\text{carbon dioxide} + \text{water} \xrightarrow{\text{sunlight}} \text{glucose} + \text{oxygen}$$

$$6CO_2 + 6H_2O \xrightarrow{\text{sunlight}} C_6H_{12}O_6 + 6O_2$$

Glucose is the basic fuel as has already been mentioned, and, what is more, this glucose can be used as a substance to which other molecules can be added, or in which parts can be

4

substituted to produce more complex sugars or proteins. The latter require nitrogen and phosphorus, and these, plus other elements necessary in small amounts, are obtained in solution from the soil in the form of simple salts, for example potassium nitrate (KNO_3), magnesium phosphate ($Mg_3(PO_4)_2$). Green plants are thus the food factories of the living world; they are solar energy traps 'par excellence', collecting the sun's radiant energy and storing it up in complex foodstuffs of their own making, to be used by themselves, or, in time, by other organisms which feed on plants. They also do the important job of refurbishing supplies of oxygen. The Biblical saying 'All flesh is grass' has been quoted many a time by biologists trying to emphasise the all-important function of photosynthesis. It does indeed hit the nail on the head, for all living things are entirely dependent on the 'greenery' be it of meadow or lake, which so obligingly provides the component parts and running fuels for us all.

THE CELLULAR STRUCTURE OF LIVING ORGANISMS

Another feature common to most living organisms is that their bodies are built up of one or many units called cells. All cells have a certain basic characteristic structure. If you looked at a cell under an ordinary light microscope, you would see such characteristic features as an outer cell membrane, which may be a thin layer of proteins and fats as in animal cells, or may be supplemented by an outer wall of thick cellulose as in plants.

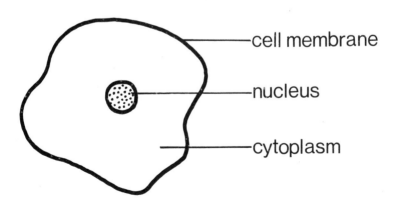

Fig 2 Here are shown the features of an animal cell as seen under the light microscope, X800.

Within this is a mass of an apparent jelly-like material the cytoplasm and lying in the cytoplasm is a smaller 'blob' which if stained with a dye usually shows up darker than the rest of the cytoplasm. This is the nucleus. Simple cells can be obtained by scraping the inside of the cheek with a match stick and examining them on a slide under a microscope. These cells are approximately 1/20 mm in diameter.

The business of life goes on in the cytoplasm. The nucleus, which contains the very special substance DNA (deoxyribonucleic acid) is the controlling centre from which information and instructions are passed to the cytoplasm.

It is in each cell that the machinery of life operates; indeed study by biochemical techniques and electron microscopy reveals that each cell is a complex system of bits of machinery. There are many microscopic organisms which consist of only one cell but most of the plants and animals with which we are familiar are vast assemblages of millions of cells, some cells specialised for particular functions, others carrying out a variety of jobs. Our organism is not really just a machine but an incredibly complex system of machinery. So, one of the ways in which the angler can now think of his fish is as a system for releasing the energy in the food provided by plants and converting it into energy for the business of life, for watching or waiting, holding station against current, swimming, darting for the bait, putting up a good fight, *etc.* He is an assemblage of tiny cells all operating in co-ordination as parts of an overall system and each containing the essential machinery for the living processes.

But in saying all this one has only scratched at the surface of a complete explanation of what is a living organism. Our fish is an elusive fellow and though we may have found out a good deal about how he ticks, there are sides to him which still remain shrouded in mystery. In the last section of this chapter we shall make him reveal a little more of his structure and functioning, but first some attention must be paid to his relatives in the plant and animal kingdoms.

DIFFERENT KINDS OF LIVING THINGS AND SOME WAYS OF NAMING THEM

Mangroves and man, mushrooms and microbes may seem to have little in common, and yet the principles just discussed apply to all of them. Clearly, they do things in rather different ways. How have they all come to be as they are, and where in the scheme of things does our fish fit in? By comparing different types of living things from evidence left by fossils, from the distribution of living things over the globe, from the study of genetics and inheritance and from experiments with living plants and animals, there has emerged a theory to explain the present situation. This is the theory of evolution by natural selection. It was first expounded in its simplest form by Charles Darwin, and still stands in a rather more sophisticated form with a great deal of evidence in support of it. It is thought that life originated gradually on our planet from non-living chemical substances in solution in the waters of ancient seas. The first crucial stage was the growth of enormously large and complex molecules. Amongst the tremendous variety of these substances, some were more stable than others and some had chemical properties which made them able to reproduce themselves. These would survive, and others, not so endowed, would eventually disintegrate again into simpler substances. Further developments were then probably the aggregation of these large stable molecules into 'combines', and the acquisition of their ability to tap the sun's energy in synthesising fuel and building materials (photosynthesis) and to utilise the fuel by 'burning' it in oxygen (respiration) for supplying energy for other activities. Just what the first living aggregations were like is a matter for speculation. There were probably very many different kinds. Some were better fitted for survival and self-reproduction than others and were selected to continue their existence. By chance chemical happenings, within the protein molecules of the cytoplasm or, in particular, within the nucleic acid molecules of the nucleus, variations in these primitive living masses would occur. From the variants, again, the best fitted for survival would be selected and the less well fitted would succumb to the constant tendency of matter towards disintegration and would be dissipated into the general surrounding 'soup'.

From our primaeval soup we must now take an enormous leap to the present day, landing, one hopes, on solid ground. Space only permits to say that the theory of evolution by natural selection suggests that from these primitive aggregates of proteins, nucleic acids, fats, carbohydrates, the vast array of present-day living forms have come into being. This has happened simply by the occurrence of variations on existing themes followed by the selection of the varieties best adapted to survive under prevailing conditions, and the inevitable 'weeding out' of the less well-adapted varieties.

The chart (Figure 3) gives a picture of the course such evolution is thought to have taken over the millions of years that life has been in existence. It is a sort of family tree of the whole of the plant and animal kingdoms.

Naming and describing living things

When an ecologist goes out to study things in their natural environment there are various descriptive words he will want to use. So, let us now set out a brief glossary of some of these.

Species – the smallest group in classification and the finest point (apart from sub-species which is sometimes used within a species) that one gets to in naming a plant or animal. Individuals belonging to a species are more like each other and more closely related to each other than to members of any other species. A species is an interbreeding group of living organisms, the members of which cannot normally breed successfully with members of any other species. (Many species have common names handed down by tradition. There are, however, many that do not have common names and moreover there may be local variations which cause confusion. Scientific naming of species is an attempt to overcome these loopholes and variations. Every species has, as its complete name, a genus name (see next paragraph) and a species name.)

Genus – a group of related species more like each other
(plural genera) and more closely related to each other than to those of any other group of species. The genus is, therefore, the second smallest group in classification. Any individual plant or animal is given a double name (genus and species). For example the goldfish is Carassius (Genus) auratus (species).

Family,– a family is a group of related genera and an
Order, Class, order a group of related families and so on. In
Phylum, etc. the tree diagram (Figure 3) most of the groups shown are either phyla (plural of phylum) or classes, for example phylum Annelida, class Polychaeta.

Biotope – a type of environment for living things, *eg* a

8

	stone wall, a stretch of clay soil, a riffle in a stream, or a pond.
Population –	a group of individuals of a species living together in a biotope (sometimes termed a 'species population').
Habitat –	the particular conditions of life for a particular species population *eg* the usual habitat of trout is a swiftly flowing stream or river.
Community –	a group of populations of plants or animals living together in a biotope, *eg* the plants and animals of an oak woodland can be said to be a community, likewise the plants and animals of a lake or stream. Sometimes there may be several communities within a larger community, *eg* one can talk about both the shore community and the open water community of a lake.
Ecosystem –	a system of living and non-living things considered together, *ie* a community and its environment all together, constitute an ecosystem. *Eg* the plants and animals plus the physical and chemical features of a lake, considered all together and interacting with each other, constitute the lake ecosystem.

Identifying plants and animals (and fish in particular)

All living things known to scientists have been classified, that is put into a species pigeon-hole within a genus block of pigeon-holes within a family and so on. Each plant or animal known to man is named by its genus name and its species name. Thus the goldfish is named *Carrassius auratus*, we ourselves *Homo sapiens*. The genus name always has a capital initial letter and the species name a small letter.

It would be impossible here to go into any detail over all this. There are, however, numerous books designed to help one to identify any plant or animal one may come across. A list of some useful ones is given at the end of this chapter.

A few brief words about the naming and classification of the fishes. Our freshwater fishes belong between them to 10 different orders of the class Pisces, sub-class Teleostei. A good summary of the names of the orders, the names of the families within them, and the genus and species names of the fishes in these families, together with their common names, is given at the beginning of 'The Observer's Book of Freshwater Fishes' (revised

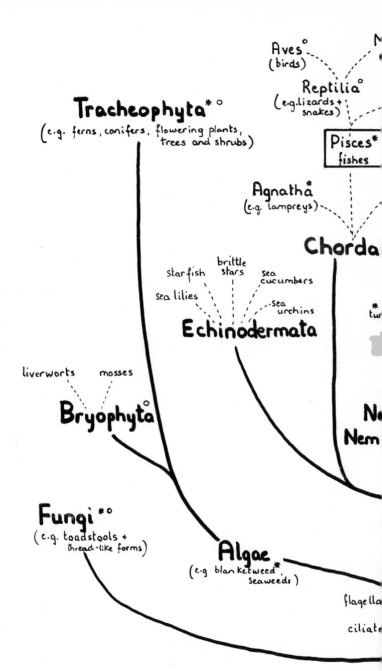

Fig 3 A simplified version of an evolutionary tree of the plant and animal kingdoms. This symbol * indicates representatives in freshwater and this symbol ○ shows representatives that are

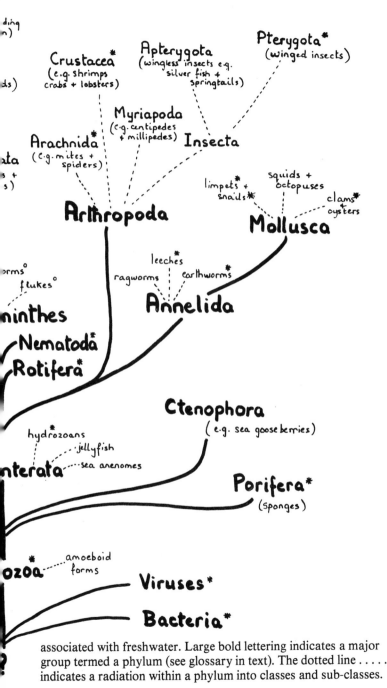

associated with freshwater. Large bold lettering indicates a major group termed a phylum (see glossary in text). The dotted line . . . . . indicates a radiation within a phylum into classes and sub-classes.

edition) by T B Bagenal. For accurate identification of individual fish the book by P S Maitland is a good one. It is laid out in key form (as indeed are many of these identification books). This means that it works by asking you pairs of questions about your specimen. For example the first pair of questions might be

1 Round, with purple spots and smooth edges. .... 3.
2 Square, with stars and stripes and frilly edges. .... 5.

If your specimen corresponds with the first description you go on to another set of questions designated by 3. If it corresponds with 2 you go on to 5 and so on, until eventually you come to a stop at a genus and species name. That, with any luck, should be the name of your specimen! Many happy hours can be spent 'running down' a plant or animal whose name is unknown to you by means of a key of this sort.

THE FISH AS A LIVING ORGANISM

Figures 4 and 5 show how a typical teleost fish, a sort of average freshwater angler's fish, is constructed, from outside and inside. You will note that the cells we talked about earlier are too small to show in such a diagram, but what one can show is the way the cells are grouped together to form organs, for example the heart, the intestine *etc,* which carry out particular functions. Each of these functions we shall take very briefly in turn, and describe how the organs concerned carry out the jobs for which they are adapted. Our knowledge of the workings of the fish body has accumulated as the result of many years of patient dissection and experimentation by numerous

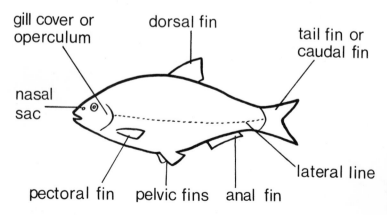

Fig 4 This diagram shows the external features of a typical teleost fish.

12

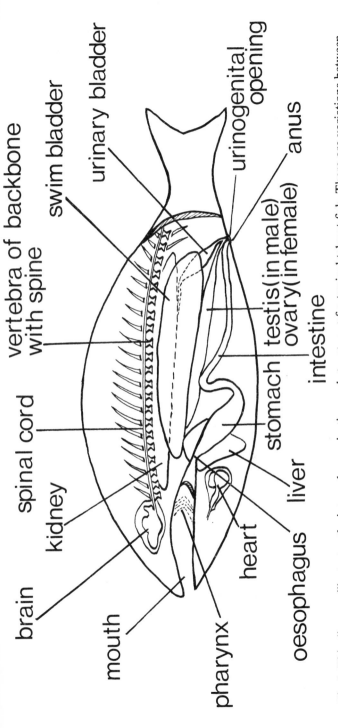

Fig 5 This diagram illustrates the internal organisation and structure of a typical teleost fish. There are variations between different species in the structure of the digestive system (*eg*, plant-eating cyprinids lack a stomach) and in many of our freshwater fishes there is a connection between the swim bladder and the back of the pharynx. Otherwise, the basic organisation is similar for all species.

13

biologists from all over the world. What follows then is just a brief summary of what we now know about the internal functioning of a fish. If you wish to delve further into the matter there are suggestions for further reading at the end of the book.

(i) Respiration

The actual burning up of glucose in the presence of oxygen (see page 3) takes place all over the body, in all cells, and particularly in the muscle cells which need a lot of energy. One problem which the fish has to deal with then is getting oxygen from the surrounding water to the cells. This is accomplished with the aid of the gills and the blood system (see diagrams Figures 6 and 7). The gills lie under the protective gill-cover or

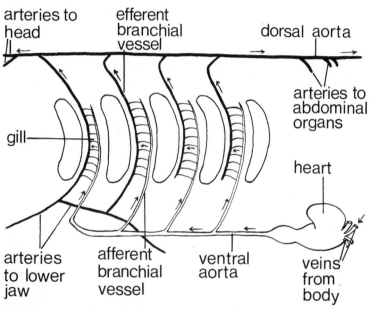

Fig 6 This shows the arrangement of the blood vessels and the circulation of the blood in the gill region. The diagram represents a side view of one set of gills. There is a single ventral aorta and a single dorsal aorta, but the afferent and efferent vessels taking blood to and from the gills are paired so that there is one set for the left-hand set of gills and another set for the right. This diagram shows only the blood vessels for the left-hand set of gills. The arrows show the direction of blood flow. Blood vessels drawn ══ contain de-oxygenated blood returning from the body to the gills. Blood vessels drawn with heavy lines contain oxygenated blood.

14

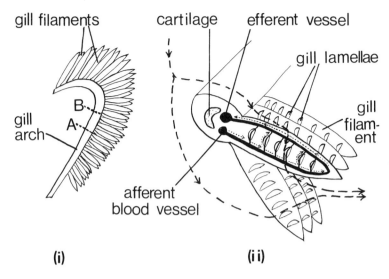

gill filaments   cartilage   efferent vessel

gill lamellae

gill arch

B.
A.

gill filament

afferent blood vessel

(i)   **(ii)**

Fig 7 Illustrates the structure and functioning of the gills of a teleost fish. (i) A single gill drawn to show the arrangement of the gill filaments on the supporting arch. (ii) Block diagram of a section taken between A and B to show the internal and detailed structure. Each filament has numerous tiny gill lamellae, which are flap-like projections from the surface. (Only a few are shown here.) Note the counterflow of water and blood. – – – – – – = water flow . . . . . = blood flow. (Redrawn from Bijtel in Brown, Physiology of Fishes, 1957.)

operculum. The fish takes water through its mouth and sends it out past the gills and under the edge of the operculum. The gills are fine feathery structures very well supplied with tiny blood vessels (minute tubes called capillaries) which are very near the surface *ie* near to the water, and contain blood pumped from the heart and flowing in the opposite direction to the water. Oxygen dissolved in the water is picked up by the blood and carried off in the continuous flow to larger blood vessels and thence to the main arteries leading to the body. As oxygen is snatched away by the blood (which incidentally contains a special red substance called haemoglobin which has an affinity for oxygen) more passes from the water through the surfaces of the gill filaments, and into the next bit of blood to come along, so to speak. The nett result is a constant diffusion of oxygen from water to blood, whence it is carried off to supply the body. The thing of great interest about this system is the enormous surface area of the gills (see Figure 6) and the efficient counterflow

15

system of blood and water, making it possible for up to 88% of the oxygen in the water to be absorbed.

(ii) Feeding

Food is taken in at the mouth, but the problem is – how to get the 'goodness' from a chewed up shrimp or bloodworm to the muscles at the end of the tail. The answer is a specially evolved digestive system (see Figure 5) working in conjunction with the blood system. A tube from the mouth leads to the stomach where food can be further churned up into finer particles. Here, and in the intestine, the tube leading from the stomach, juices containing digestive enzymes are added to the food and it is gradually broken down into simple substances which can pass through the lining of the intestine and into the blood. (Some cyprinid fishes lack a stomach and the food simply passes through a long coiled intestine in which a gradual process of digestion takes place). Enzymes are substances which help along and speed up (or sometimes slow down) chemical reactions taking place in the body, and the digestive enzymes help in the process of splitting up the large molecules of proteins, starches and sugars (carbohydrates) and fats in the food into smaller units which can pass easily into the blood.

Once in the blood, the food materials, now in a simpler form, are whisked off to different parts of the body. Much of the blood, on its way from the intestine, passes through the liver where some of the food materials are processed and others stored. Inevitably a certain amount of water is taken in with the food and absorbed into the blood, but a freshwater fish does not deliberately drink water – the reasons for this will be explained in Chapter 2. Anything which is not absorbed, mainly solid matter which is either difficult to break down or contains substances which cannot be absorbed, is got rid of at the end of the intestine which opens at the anus somewhere between the pelvic fins and the tail.

We have talked about food being taken in and either being absorbed or got rid of. But how about the shark which was found with an anchor in its stomach – not to mention all those hooks found in freshwater coarse fishes? It is perfectly possible for a metal object to be swallowed and to get lodged in part of the digestive system. In the stomach the digestive juices contain hydrochloric acid – fairly strong corrosive acid – which would in time dissolve away most metals. Apart from what I should think is considerable initial discomfort, it is possible for fish to come to little harm through swallowing a metal object as long as any wound does not become infected. One does not of course know what might be the long term effects of absorbing small amounts of metals some of which might be toxic ones such as nickel and cadmium. The shark, however, – so I am told – had survived its internal anchor for some time, judging by the state of corrosion of the metal.

(iii) Movement

Swimming is accomplished by muscles acting in conjunction with the firm bony framework of the skeleton. Muscles are made up of millions of tiny fibres – each fibre being capable of rapidly contracting in length. When all the fibres in a piece of muscle contract at once a tremendous pull is exerted on the skeletal parts to which the muscle is attached, the body moves, and a force is exerted against the water. In the fish the muscles at the tail end of the body and in the tail itself are the important ones for swimming. When the muscles on one side contract, the tail with its fin is curved towards that side, pushes against the water, and the fish is forced forward. The muscles on the other side then contract pulling the tail the other way, and leaving the muscles on the opposite side to be passively stretched. This alternating action of the muscles and the flicking of the tail from side to side effectively sculls the fish through the water. Other fins are used as stabilisers and brakes.

(iv) Excretion

As a result of all the activities of the body, involving many complex chemical reactions, various waste materials accumulate and have to be got rid of before they have a poisoning effect. This, together with getting rid of excess water, is the job of the kidney which is situated above the swim-bladder, closely adherent to the body wall and muscles of the back, and not, as in mammals, 'kidney shaped'. In the simplest possible terms, blood is constantly kept flowing through the kidney and it acts as a sort of filter, sifting out the contents of the blood which need to be got rid of, and leaving the rest. A solution of waste substances, the urine, is sent down tiny tubes into two main tubes and thence to a urinary bladder where the urine can be temporarily stored. It is eventually got rid of to the outside world through the opening common to the excretory and reproductive systems, the urino-genital opening, just behind the anus.

17

(v) Reproduction

All vertebrate animals reproduce sexually, that is they have some sort of 'sex-life' which is an adaptation for producing offspring. This basically involves the production of eggs by the female and sperm by the male. The eggs (spawn) are produced in the ovary (hard roe) of the female and the sperm (milt) in the testes (soft roe) of the male. Eggs and sperm are brought together so that a sperm cell can fuse with an egg cell and the egg thereby made capable of developing into a new individual. This process of the fusion of the sperm with the egg is called fertilisation. A single fish sperm cell is of the order of 1/500 mm long and it can swim about by means of a thread-like tail. Fortunately (perhaps) for fish, sperms are reasonably happy swimming in the water of a river or pool for a short while. This means that fish are not presented with the problems which land animals tend to have in conveying the sperms to the eggs. In other words, the sex lives of fish do not need to be quite as complicated as those of land-dwelling animals. Basically all that needs to be done is for the eggs to be deposited in the water and the milt on top of them. In practice some fish do go in for protecting their eggs and ensuring that fertilisation is safely accomplished. Also the developing fry are kept hidden from hungry passers-by. Examples of fish which do protect their eggs to some extent are the salmonids and the stickleback.

Although the process of fertilisation of the eggs by the shedding of milt on to them is basically the same in all our freshwater fishes, there are differences in mating behaviour and in the subsequent development of the eggs and fry. Quite a lot is known about the reproductive habits of the salmon and trout and the nest building antics of the stickleback, but very little is yet known in detail about the coarse fish. Research in the future may well unearth unsuspected subtleties in the ways our coarse fishes are adapted for producing offspring. A little more about what is known of the breeding of coarse fishes appears in Chapter 4.

(vi) Hormones

These are the body's chemical 'messenger boys'. They are substances produced by special organs, the glands (for example the pituitary gland in the brain, the thyroid in the neck region). In mammals the hormones have been shown to help in controlling bodily reactions – particularly the rate of living (that is whether one is highly active or rather sluggish) and rates of growth. Hormones work in close conjunction with the nervous system, and, for example, may come into play in situations of potential danger to which the nervous system has been alerted. Thus the sense organs and nervous system may detect a dangerous situation, the nervous system may then stimulate a gland to secrete a hormone which is carried to the muscles and may set them contracting more rapidly, set the heart beating more rapidly thus speeding up the supply of oxygen and fuel to the muscles and so on.

Although far more is known about mammalian hormones than fish hormones, fish do have similar glands to mammals and it is thought that the hormones they produce are similar and work in the same sort of way. This is still a wide open field in fish physiology, however, and future research may throw up some interesting things about fish hormones.

(vii) Growth

Looking around at the living world one can observe that living organisms always grow bigger. Weight may vary up or down (as some of us know to our cost!) but increase in length is irreversible (except for cases of shrinkage of the backbone in old age and odd cases in the animal world connected with a change in state for example from caterpillar to butterfly).

Growth occurs by the production of new cells, and for this a good supply of food is of course needed, over and above the quantity needed for everyday 'ticking over'. In other words a supply of food is needed over and above the maintenance requirements.

There are probably many factors which affect growth, both internal and external. Internally the production of hormones by the pituitary and thyroid glands affects and controls growth. Outside factors, however, affect either the growing parts themselves or the glands which secrete the growth controlling hormones. Among such factors are the quality of the food, temperature, light and the degree of crowding.

Many experiments have been done to try and find out just what conditions are best for promoting rapid and constant growth. And of course this research is of great importance for anglers who want to be able to know how best to cultivate specimen fish. An example of some experimental work was that carried out by M Brown (now Mrs M Varley) in the 1940's. Her experiments were on the effects of temperature on the growth of brown trout. She kept two-year-old trout in aquaria, feeding them all regularly and well, but keeping the tanks at different temperatures. She found that the fish grew best between 7°C (44·6°F) and 9°C (48·2°F) and between 16°C (60·8°F) and 19°C (66·2°F). This can probably be explained in terms of the balance between appetite (probably controlled by hormones), activity and maintenance requirements. Below 7°C (44·6°F) the fish are fairly sluggish, and feed little. At 7°C appetite and activity begin to increase giving the observed increased growth rate between 7°C (44·6°F) and 9°C (48·2°F). Above 9°C the maintenance requirements suddenly shoot up. To begin with increase in appetite does not quite keep pace but at about 16°C (60·8°F) the maintenance requirements are increasing less rapidly, appetite has increased to a high level and growth is good. At about 19°C (66·2°F) appetite appears to fall off, while the maintenance requirements stay the same, so that growth is then not so good. Other experiments with brown trout have given slightly different results – the situation is obviously not a simple one. But it does appear that temperature has a considerable influence on growth in trout and doubtless in other species as well.

In connection with the effects of temperature on growth, studies are at present being made of the effects of heated effluents from power stations. Some work by D Cragg Hine on a population of roach in such an effluent near Peterborough has shown that the young roach grow faster in the early stages but that this later slows down so that the overall growth rate is the same as that in the main part of the river. It is possible of course that the levelling off is due to excessive competition for food in the heated channel.

Another study by D J A Brown has shown that the fry of several species of coarse fish have grown faster in the first year of their life downstream of a power station effluent and that the growing season is extended one month or more beyond September when growth normally ceases. It would be interesting to try and find out whether there are any conditions downstream of the entry of heated water where this effect is carried through into adult life.

In general it is important to remember when thinking about the growth of fish that there is no single fixed adult size for any one species. One comes across stunted fish and fast-growing fish and most of the evidence to date seems to suggest that the differences are due to factors in the environment rather than to hereditary factors.

(viii) Sense Organs and Nervous System

The ways in which a fish finds out about what is going on around it, and the ways in which it reacts, will be discussed in a separate chapter (Chapter 5).

Suggestions for further reading on classification and identification

R Freeman. Classification of the animal Kingdom. English Universities Press and Reader's Digest Association.

T Bagenal. Identification of British Fishes. Hulton.

T Bagenal. Freshwater Fishes. Observer Series, Warne Bros.

J Clegg. Pond Life. Observer Series, Warne Bros.

T Macan. Guide to Freshwater Invertebrates. Longmans.

P Maitland. Key to British Freshwater Fishes. Freshwater Biological Association, Scientific Publications.

H Mellanby. Animal Life in Freshwater. Methuen.

Harvey Torbett. Angler's Freshwater Fishes. Putnam.

B Muus and P Dahlstrom. Guide to the Freshwater Fishes of Britain and Europe. Collins.

Clapham Tutin and Warburg. The Flora of the British Isles. Cambridge University Press.

Keble Martin. The Concise British Flora in Colour. Sphere Books Ltd.

Chapter 2 – Lakes and Rivers

THE WATER – AND DO FISH DRINK IT?

It is now time to prepare the ground for studying our plants and animals in their natural surroundings. First, let us consider this colourless, odourless, innocuous liquid water. If you were to analyse living cells you would find that water makes up 65–75% of their weight. The researches of chemists have shown us that water itself consists of tiny particles (molecules) and that each molecule is made up of yet smaller particles (atoms) of hydrogen (H) and oxygen (O). In each molecule of water there are two atoms of hydrogen bound to one of oxygen. In water in its normal liquid state some of the molecules are split up into electrically charged particles called ions – hydrogen ions (Positively charged and written H^+) and hydroxyl ions (negatively charged OH^-). Now, for reasons connected with the arrangement of the atoms in the water molecule it happens that water is a very good solvent for other substances; that is molecules and ions of other substances disperse easily and stably between the molecules and ions of water. Pour common salt into a cup of water and the salt itself disappears – the salt particles have dispersed in the water and a solution has been formed. It is in solution in water that chemicals are transported around inside living organisms and it is solution in water that all the living processes inside cells are carried out.

Now one might well think from all this that to actually live in water would be most convenient. Certainly, there is no danger of losing water by evaporation into the air, which is a problem for land-dwelling organisms. Watery solutions, however, have a property which presents a particular problem to water-dwellers. You can observe this quite easily by a simple experiment with a potato, a saucer, some sugar and some water. Proceed as follows: cut the potato in half and scoop out a small hollow in the rounded end. Place the flat, cut surface in a saucer of tap water and into the hollow put half a teaspoonful of a

22

Before   **After**

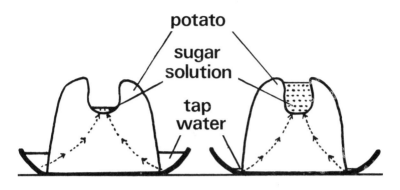

potato

sugar
solution

tap
water

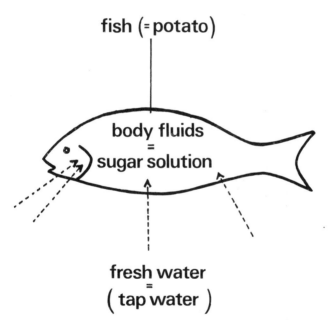

fish (= potato)

body fluids
=
sugar solution

fresh water
=
(tap water)

Fig 8 Diagram to illustrate the potato experiment which
demonstrates that freshwater fish absorb water (see text). Marine fish,
on the other hand, need to retain water to prevent themselves
becoming dehydrated. Some fish are specifically specialised in this
respect. Goldfish for instance, cannot stand salt water and will die in it.
On the other hand salmon, sea trout and eels have developed the
capacity of 'bailing out' while in freshwater and of conserving water
while in sea water.

23

concentrated sugar solution (say $\frac{1}{2}$ teaspoonful of sugar dissolved in 1 teaspoonful of water). After a few hours you will notice that the level of solution in the hollow has risen and the level in the saucer has dropped slightly. One can therefore conclude that water has passed from the saucer through the potato and into the sugar solution. Now this is what is inclined to happen in all living organisms which have any thin-walled cells in contact with fresh water. Compare the water in the saucer with the water a fish is living in, the potato with the cell membranes of the outer parts of the fish – in particular the gills, and the concentrated sugar solution with the cell and body fluids of the fish. Where a dilute solution is separated from a more concentrated one by membranes permeable to water, but not so readily permeable to substances with larger molecules, water will always pass from the more dilute to the more concentrated solution as inevitably as water always runs downhill.

For our freshwater fish, then, the consequences are obvious – water will constantly tend to pass from its surroundings into the fish's tissues, and it is in constant danger of being 'flooded out'. We can now answer the question as to whether a fish drinks. The answer is that freshwater fish do not drink in the sense that we do. They may well take in a certain amount of water with their food, but they would do better without it. The creatures which live in sea-water have the opposite problem to cope with. Their body fluids are slightly less concentrated than their surroundings and consequently they tend to be in danger of 'drying out'.

Then how do fish cope with the problem? Part of the answer lies in the possession of a covering of relatively impermeable skin and scales. But there always remain a few unprotected surfaces through which water can pass freely, for example the gills. Freshwater fish therefore have to have some means of 'bailing out' and this they do through their kidneys. Marine fish need the capacity to retain water, and so prevent themselves becoming dehydrated.

Some species of fish are totally specialised for the freshwater life within fairly narrow limits. If, for example, you popped your goldfish into a bowl of sea-water it would soon show signs of distress and would eventually die, ironically enough from lack of water. Used to constant 'bailing out' it cannot immediately adapt to do the opposite. Members of the carp family are specialised for life in freshwater only (they are said to be stenohaline) but there are certain freshwater organisms which are tolerant of a wide range of salt concentrations (euryhaline). Notable examples are certain species of algae, shrimps and, amongst fish, the stickleback.

Truly remarkable, however, are the fish which can pass from sea to freshwater and vice versa at different stages of their life histories. The migratory salmon, sea trout and eels have gained an independence over the restrictions which this aspect of life in water can impose – they can 'bail out' while in freshwater and conserve water while at sea and so they form a link between the ecosystems of inland waters and the sea.

WATER DENSITY AND SUPPORT
The density of water is of considerable significance in the lives of aquatic organisms. In comparison with air, water is an idle paradise in which to loll at ease. Very little energy and materials need to go towards support in this relatively dense medium. On the other hand, it is harder to propel oneself through water, and, what is more, turbulent water exerts stronger forces than turbulent air. The former factor particularly affects animals which seek their food in the open water. The latter factor imposes considerable limits on the possibility of a livelihood in fast running streams and rivers. We shall return to this later in Chapter 2 when considering the physical characteristics of streams and rivers. Density of the water is of course affected by temperature – the warmer it is the less dense it becomes. But this also will be discussed later under the heading of temperature.

OXYGEN – 'SUMMER KILLS' AND 'WINTER KILLS' AND THE USE OF KEEP-NETS
We have already touched on the subject of the presence of dissolved oxygen in the water and have discussed the need for a good supply of oxygen for respiration. Now to consider the question of oxygen in more detail. We have spoken of substances such as salt and sugar dissolving in water. In similar fashion gases dissolve in water – the molecules of the gas dispersing between the molecules and ions of water. Oxygen does just this and oxygen molecules pass into and out of solution at the air/water surface. But the amount which is able to pass into solution is affected by temperature – if you measured chemically the amount of oxygen in a certain volume of water at a variety of temperatures your results would show that as the temperature of the water fell the amount of oxygen

dissolved in it increased. All other things being equal, water is more highly oxygenated when it is near freezing point than when it is luke warm.

An experiment to show the relationship between temperature and the oxygen content of water

Having already observed that fish need oxygen in order to survive, and that they acquire it by passing water over their gills, a simple experiment with a goldfish will help to show this relationship between temperature and the oxygen content of water. The apparatus required is as follows: one goldfish (which will come to no harm), one two-pint saucepan, one bottle with a well fitting screw top (not more than two pints capacity), two glass dishes (*eg* pyrex mixing bowls) and the use of the stove. Fill the pan with water and slowly heat it to just below boiling point. Meanwhile, warm the bottle in the oven set at about 200°F. After keeping the water at just below boiling point for a few minutes pour it gently into the bottle (which one hopes will not crack having also been warmed up). Screw the bottle top on immediately so that the bottle is completely sealed and allow it to cool to room temperature (which will take about $\frac{1}{2}$ hour). Into one of the bowls place water from the tap, and, when this is at room temperature, place the goldfish in it and observe the rate of its respiratory movements (mouth and operculum).

Next pour the water from the bottle very gently into the other bowl and transfer the goldfish to it. Again watch carefully to see what effect this has on the fish (and then put it back into its other bowl as soon as you can).

You will find that in the water which had previously been heated up the goldfish soon showed signs of distress and its breathing movements became more rapid, that is, it opened its mouth more frequently and stepped up the rate of taking water over its gills. It seems likely that it is doing this in an attempt to get more oxygen to its gills and we can infer that the oxygen content of this water is lower than that of the fresh tap water. This experiment has thus helped to confirm that raising the temperature of water does lower its oxygen content.

One must remember, however, that if one left this bowl of de-oxygenated water open to the air at room temperature its oxygen content would eventually become the same as that of the other bowl. But this would take some time. Oxygen diffuses much more slowly in water than it does in air and it takes de-oxygenated water much longer to become re-saturated with

26

oxygen than it does to freshen up a stuffy room by opening the window.

Let us now turn to the dangers of oxygen depletion in pools and lakes. Oxygen is put into the water at the surface by diffusion from the air and also it is supplied by green plants when they are carrying out photosynthesis during the day. It is removed when the temperature is raised, for example in shallow water on a hot summer's day, and also by the respiratory activities of the living organisms in the water. The dangers of severe deoxygenation in a mountain stream are non-existent because the temperature is always low, there are usually fairly few plants and animals living in it, and its turbulence increases the surface in contact with air, mixes it thoroughly and keeps up a high level of aeration.

It is a different matter altogether in a lowland pool, rich in plant and animal life and shallow enough for all the water to be warmed up considerably in the summer. Remember here that in sunlight the plant life in the water will be carrying out photosynthesis and in the process will be producing oxygen, so during the day the oxygen which is used up by the respiration of living organisms can be replaced to a large extent by the plants. At night, however, both plants and animals carry on using oxygen for continuous basic respiration while photosynthesis ceases. On a warm summer's night the minimum of oxygen is held by the water and only a little is dissolved at the surface. It can happen, therefore, that it is being used up by plants and animals faster than it is being replaced. Add to this the oxygen requirements of bacteria and fungi getting busy on the season's dead organic material on the bottom and one has a crisis situation which may well result in severe lack of oxygen. These are the conditions under which a 'summer kill' of fish may occur, and they happen typically between midnight and dawn.

Even during the day it is possible to have localised regions of deoxygenation particularly in shallow places where there are few plants. This brings us to the question of the use of keep-nets. When choosing a place for his net the angler is likely to go for a spot near the edge where there are no plants for it to get tangled up in. This is precisely the sort of situation which in hot summer weather will suffer from lack of oxygen – little oxygen dissolving in the water, no plants to supply it and the fish in the net using up what little there is. The shortage of oxygen is bound to cause the fish distress and, though it may not die, it will certainly return to the water in a weakened condition. The moral of this is, of course, to choose a shaded cool place for one's keep-net, preferably where there are some plants.

Consider now a day in winter with the pool frozen and covered with snow. Here two factors operate – there is no redissolving of oxygen at the surface, and there is little light for the aquatic plants. Oxygen used up cannot be replenished and the conditions leading to 'winter kill' are created. A practical solution for avoiding winter kill is to keep ice-covered stretches of water if possible free of snow by brushing or applying lamp-black. The light to some extent can penetrate snow-free ice and some of the algae and submerged rooted plants may be able to continue to photosynthesise.

No mention has been made so far of carbon dioxide dissolved in the water. If a plentiful supply of oxygen is necessary, so indeed is a supply of carbon dioxide for use by plants in photosynthesis. But whereas oxygen is used both day and night by both plants and animals, carbon dioxide is used only during the day and by plants only. What is more, carbon dioxide dissolves more readily in water than does oxygen, and it is constantly being produced in respiration. There is therefore little danger of carbon dioxide shortage – rather at night it is possible that there may be too much.

We can see therefore that as far as dissolved gases are concerned it is the supply of dissolved oxygen which is one of the crucial factors in freshwater ecology, and is of great importance in the lives of fishes.

OTHER SUBSTANCES IN SOLUTION – HARD AND SOFT WATER

Theoretically it should be possible to take a seedling from the garden and to keep it alive (albeit at the same size) indefinitely in a jar of distilled water on your kitchen windowsill. It will be supplied with water, carbon dioxide and sunlight, and should be able to manufacture its own food, which it can then use as a source of energy to keep it 'ticking over'. But, as every gardener knows, though you may be able to keep your seedling 'ticking over' for some time it will not grow into a large healthy plant. To do this it needs other substances which it can combine with the carbohydrates manufactured in photosynthesis to produce the proteins necessary for growth and repair. If you have some spare seedlings from your garden, and could persuade a chemist to supply you with a variety of simple water-soluble salts, you could set up a simple experiment to test out which substances have an effect on the growth of your seedlings. I will leave that to your imagination to design – (if simple salts, such as potassium nitrate, ferric chloride, sodium sulphate, are unobtainable then use proprietry brands of plant fertiliser which will have the ingredients marked on the packet).

The object of this exercise is simply to demonstrate that there are elements other than carbon, hydrogen and oxygen which are essential for the healthy growth of organisms. The ones which have been shown to be important are nitrogen, phosphorus, silicon (for certain algae), potassium, sodium, calcium, magnesium, iron and certain other elements in minute quantities.

The presence of these essential elements in water depends largely on whether they are supplied by erosion of rocks and soil at the sources of streams or over which streams and rivers flow. Water which originates from peaty uplands already heavily leached of salts tends to contain very low concentrations of dissolved substances. Lacking in particular from this type of water are the calcium and magnesium carbonates and sulphates. Such water is said to be 'soft' and is usually rather acid. Another characteristic is that it has a low electrical conductivity – mentioned here because it is one of the measurements ecologists use to compare different waters. 'Hard' water, on the other hand, contains large amounts of dissolved salts, is alkaline and has a high electrical conductivity. It is the soft water that makes the whisky and the hard water the beer, but when it comes to freshwater life there is no doubt that hard water can support a richer flora and fauna than can soft water, simply because it contains a better supply of the essential building materials. In addition there is some evidence that acid water may have a directly harmful effect on some organisms, including fish.

Here mention must be made of methods of measuring hardness and softness – one of the freshwater ecologist's favourite occupations! The degree of acidity or alkalinity is measured on a scale of units called pH. The pH value is a measure of the number of hydrogen ions in a given volume of water – the more hydrogen ions there are the more acid the water. But such is the mathematics of the matter that the actual scale of pH values works inversely so that a low value indicates high hydrogen ion concentration and therefore high acidity. The scale is from 1–14 and so centres round the figure 7. A pH of 7 represents neutrality; above 7 the trend is towards alkalinity and below 7 towards acidity.

The actual measurement of pH is done either chemically with coloured substances which change colour in a characteristic way at different pH's, or by electrical means. The methods which rely on colour change are less sensitive than the electrical methods, but seem to me to be more reliable in the sense that electrical apparatus tends to need constant checking and adjusting and delicate electrodes need a great deal of care when being used frequently in the field.

Testing for pH

It might well be that an angling club might decide to test its waters to find out what pH they were. This is perfectly possible without having to resort to expert help. The simplest way of going about it is to use the method which gardeners use to test their soil. Gardening shops sell a variety of colorimetric apparatus for soil testing. Some involve placing a drop of a coloured solution into the sample of soil mixed with purified water (in our case into a sample of the pool or river water) and observing the change of colour, if any. One then compares the test solution with a colour chart provided with the kit, and one can read off the pH value from the colour on the chart with which the solution matches.

Other kits simply have pieces of paper impregnated with a chemical which changes colour with change in pH. In this case one simply dips the piece of paper into the water and compares the colour with the chart provided. One word of warning − some gardeners' kits do not go much above a pH of 7·5. This is alright for finding out about acidity, and indeed, since we want on the whole to avoid acid conditions in angling waters, this is adequate if all we want to do is to find out whether or not our water has a pH of below 7. If however one wishes to find out just how alkaline the water is then one wants a method which extends the range up to about 8·5–9. Another warning about the cheaper gardeners' kits (I have obtained one for 11p) is that they give only a rough indication in words, for example 'very sour' (very acid), 'sour', 'sweet' (alkaline) 'very sweet'. Again this is alright for a rough indication. If you want to be more accurate then more sophisticated apparatus must be obtained − one source could be from the firms which supply educational equipment to schools and colleges.

If you go around testing all sorts of waters for pH (take your kit on holiday with you!) you will find that lowland lakes which receive their water from areas rich in limestone (calcium carbonate) will almost certainly have a pH of 7·5–8 or even higher. A moorland stream or pool where the water is draining from boggy and peaty soils may be acid enough to have a pH of as low as 4 or 3.

The measurement of other substances in the water has to be done by standard methods of chemical analysis and needs laboratory equipment more complex than that for the measurement of pH.

If, on analysis, an angling water turns out to have a pH of less than 7 and has only very small quantities of other elements (*eg* less than 2 m.equiv/litre total salts) the situation can be remedied in theory very much as the gardener treats poor soil. For example the pH can be raised by adding crushed limestone and nutrient salts can be added by using fertilisers, (see Chapter 6 for recipes). It should be remembered, however, that a large pool may require an awful lot of fertiliser and it may turn out to be too expensive to be worthwhile.

TEMPERATURE – AND ALL ABOUT FISHING INTO THE THICK END OF THE WEDGE

We have already considered the effects of temperature on dissolved oxygen. It remains to think about variations in water temperature within a body of water or between different waters and their direct effects on the creatures living there.

As with many substances the density of water increases with fall in temperature. Water is remarkable, however, in that this only applies as far as 4°C (39·2°F). Below this temperature it becomes less dense. Therefore water below 4°C (39·2°F) will remain at the top and float there. Further cooling leads to the formation of ice at 0°C (32°F). In a lake in winter during a cold spell the water usually cools slowly, mixing as it does so, and it may cool until it is at a uniform temperature of 4°C (39·2°F). Ice then forms if cooling continues. (It is unlikely that cooling at the surface would be so rapid as to form a layer of ice floating on top of water warmer than 4°C (39·2°F).

The warming up of a body of water is effected by the sun's rays. But the heating effect only occurs to a depth of about 10 metres. Within this region all the heat-imparting radiation is absorbed. Here is a factor which exerts a profound effect on a body of water deeper than 10 metres. After a hard winter the water of a lake will be at a uniform 4°C (39·2°F). As the spring progresses the upper layer warms up and if there is windy

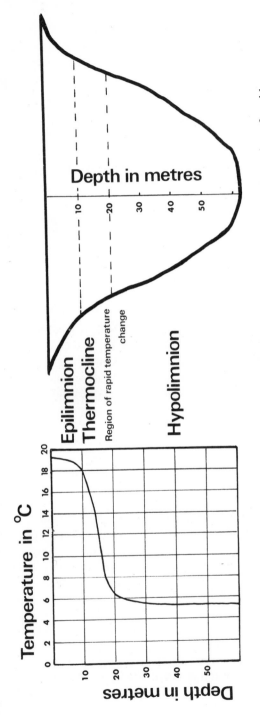

Fig 9 Diagram to illustrate stratification in a typical fairly deep lake. The thermocline marks the region of rapid temperature change. The text explains the effect of wind on the mixing of the waters.

weather there may initially be some mixing with the lower layers. Once well into summer, however, the upper layer has warmed up considerably and floats as a separate zone on top of the much colder water below. The drop in temperature from the upper warm layer to the lower cold layer is sharp and this region of falling temperature is called the thermocline. The two layers have, inevitably, been given names; the upper is called the epilimnion and the lower the hypolimnion. Under these conditions the lake is said to be stratified.

Studies of lakes over long periods involving the recording of a lot of detailed information (for example those carried out at the Windermere laboratory of the Freshwater Biological Association) have shown up several surprising things about stratified lakes. First, the stratification is maintained in summer even after very windy weather. Secondly, both the epilimnion and the hypolimnion appear to be well mixed within themselves. An imaginative experiment done in the laboratory by C Mortimer using dyed water and simulated winds using ladies' hairdriers has provided an explanation. A strong wind apparently simply displaces the epilimnion into a wedge shape with the thick end at the leeward side of the lake. As the wind ceases the wedge flows back but piles up at the opposite side under its own momentum. In this way it may oscillate for several hours, but still maintaining its separate identity. In a real lake these oscillations are called seiche movements. At the same time the hypolimnion is made to flow to and fro simultaneously and as it does so it is thought that irregularities in the bed of the lake set up eddies which produce a mixing effect. In the epilimnion mixing is thought to occur by the creation of a circular current effect – water being blown along fast at the surface and being pushed down and round in a compensatory current along the bottom of the epilimnion. As the autumn approaches cooling of the epilimnion occurs and it gradually approaches the temperature of the hypolimnion. Then finally winter gales obliterate all traces of the stratification.

Figure 9 opposite illustrates these things in a deep lake.

Before leaving the subject of seiches mention must be made of the practice of choosing the leeward side of a lake to fish on a windy day. In a stratified lake this means choosing the thick end of the wedge! Anglers have found this practice to be the most profitable. The explanation for this may simply be that food in the form of debris gets swept over to the leeward side and that the fish tend to congregate there to feed. Another explanation is that the fish prefer to keep in the warm water of the epilimnion. Remember, anyhow, that while the wind is blowing, do fish into it – and when it stops you may have to oscillate! Another consequence of stratification is the occurrence of de-oxygenation in the hypolimnion. Since there is no mixing during the summer between upper and lower layers the replacement of oxygen used up by the creatures living on the bottom is considerably slowed up. Any replacement of oxygen to the deeper parts of a lake, where no plants live, is dependent on the very slow diffusion of the gas from the epilimnion downwards and there is no mechanical mixing to aid it as there is in winter.

Shallow lakes, ponds, canals, streams, rivers etc, tend to have more uniform temperatures, and, although there may be temporary temperature gradients in shallow lakes and ponds, there is no permanent stratification. In summer temperatures may rise to about 20°C (68°F) and fall in winter to 4°C (39·2°F). As might be expected, mountain streams tend to have the lowest average temperatures while shallow lowland pools have the highest. Examples of extreme natural temperatures which have been recorded are a summer temperature of 25·5°C (77·9°F) on the river Trent in 1959 and a winter temperature of 2·2°C (36°F) on the river Test in 1960. In recent years our attention has been drawn to the artificial warming up of rivers by power station effluents. This subject comes up again in Chapter 7.

It remains for us to consider briefly the effects these temperature variations may have on living creatures. Birds and mammals are the only animals which have control over their body temperature. Most other living things have body temperatures more or less the same as their surroundings, and such creatures are called 'cold blooded'. Anyone who has ever studied elementary chemistry will remember that an essential piece of equipment is the Bunsen burner, or some other source of heat, and that heating speeds up a chemical reaction. The same applies to processes going on inside living cells and heating up a cold blooded animal speeds up its bodily processes. The converse also applies; that is, cooling slows things down. Different species of cold blooded animals tend to have different basic rates of living (called the basal metabolic rate, BMR), but all are affected by falling and rising temperature. It follows from this that some will be best suited to living in cold regions (those whose basal metabolic rate is already quite high) and others to warmer regions (those whose basal metabolic rate is

34

fairly low). In the former there is leeway for cooling before the body's processes are slowed down to a dangerous level. In the latter further warming does not speed up the metabolic rate too dangerously, that is to the point at which, for example, demands for oxygen can no longer be met. Different species of fish show interesting adaptations in this respect and in Chapter 4 we shall think about them in more detail.

LIGHT

Our essential life force, the sun's radiant energy, only penetrates to a certain depth in water, roughly little further than 10 metres. Thus no rooted plants can exist at any greater depth than this, and, moreover, photosynthesis carried out by floating algae can only occur in the epilimnion of a stratified lake. Another factor to be taken into account is that within the illuminated zone certain wavelengths penetrate further than others (the blue and green parts of the spectrum penetrate further than the red and yellow parts). This means that plants living near the bottom of the 10 metres deep illuminated zone have a different sort of light coming to them from those in the upper layers. Plants which live in deep water may be adapted to allow for this.

PHYSICAL CHARACTERISTICS OF THE BOTTOM – SUBSTRATES FOR PLANTS AND COVER FOR ANIMALS

The network of relationships between living organisms and their environment is called an ecosystem. One can set ones limits round an ecosystem very much as one chooses. For example, the whole of the land mass of England, Scotland and Wales with its flora and fauna could be regarded as one ecosystem, but within that large one a closer look would reveal several smaller ecosystems which one would be justified in considering separately. Take a lake; it can be considered as a whole ecosystem or as a set of ecosystems – the shore, the open water, *etc*. The nature and bounds of an ecosystem are often determined by its physical features.

Thinking along these lines, we shall consider some of the different physical features of stretches of fresh water, taking lakes as whole entities, and dividing rivers into reaches. Still-water ecosystems are affected by depth, size, steepness of their sides *etc*. Running waters are affected chiefly by one feature, that is the gradient, to which are related rate of flow and the

Table 1 Characteristics of some large, still bodies of water and their classification into three types. (Compiled from data in a variety of sources)

Type of water and its situation	Depth	Profile and type of bottom	Stratification	Chemical factors	Plants	Animals
Upland mountain lakes in steep-sided valleys	Very deep	Steep rocky shore subjected to wave action	Stratified in summer	Acid and often poor in dissolved salts. Rich in oxygen	Few rooted plants. Algae on stones and in plankton	Invertebrates few, both in variety of species and numbers. Fish – mainly trout, char and whitefish
Lakes in foot hills of mountains, or recently formed reservoirs in lowland areas, eg some gravelpit lakes	Moderately deep	Steep or moderately sloping shores. Stones and silty areas. Wave action where exposed	Stratified in summer (if well over 10 metres in depth)	pH usually neutral to alkaline. Moderate amounts of dissolved salts and oxygen*	Some rooted plants. Algae on stones and in plankton	Invertebrates varied and fairly abundant. Fish – mixed fauna of coarse fish and some salmonids, eg roach, perch, pike, trout
Lakes in flat lowland regions	Shallow, ie less than 10 metres	Gently sloping shores. A large proportion of sand silt or mud. Wave action where exposed	No stratification	pH alkaline. Rich in dissolved salts. May be low oxygen concentrations in summer	Abundant rooted plants of all types and both attached and floating algae	Invertebrates varied and abundant. Fish – coarse fish–in particular bream, tench and carp

* though oxygen may become depleted in hypolimnion in summer.

Table 2 Characteristics of running waters and their classification into four types. (Compiled from data in Huet and Varley)

Type of water and its situation	Temperature and chemical characters	Gradient	Rate of flow of water	Substratum and sediment	Plants	Animals
Mountain headstream	Low temperature. Acid and poor in dissolved salts. Rich in oxygen	Very steep, eg $8^0/_{00}$ or more	Very rapid and turbulent	Rocky or stony – very little sediment	Rooted plants absent. Some algae and mosses on stones	Invertebrates adapted to crawling or living attached to stones. Fish – only highly active species present, eg trout
Upland river, eg in the valleys of the Scottish highlands, Wales, the Lake District etc	Low to moderate temperatures. Fairly acid unless in a limestone area. Poor in salts. Rich in oxygen	Steep, eg $4^0/_{00}$	Rapid	Mainly stones and gravel. Some patches of sand or silt	A few submerged and emergent rooted plants. Algae etc on stones	Invertebrates as above, with some mud-burrowing species. Fish – a mixture of species but still mainly the active ones, eg trout, grayling, minnows and some dace and chub
Lowland river in rolling country	Moderate temperatures pH usually on alkaline side. Moderately rich in salts and in oxygen	Moderate, eg $1\frac{1}{2}^0/_{00}$	Moderate	Mainly sandy or silty bottom with some gravel	All types present	Invertebrates – mainly those species which live among plants or burrowing in mud. Fish – mixed coarse fish fauna with dace, chub, barbel, roach, rudd, perch and pike
Lowland river in flat country	Moderate to high temperatures. Alkaline and rich in dissolved salts. May have low oxygen concentrations	Very slight, eg $0 \cdot 25^0/_{00}$	Slow	Mainly mud	All types present	Invertebrates as above. Fish – mainly bream and tench

37

presence or absence of sediment. In very fast running streams and rivers small particles are washed away and there is of course nowhere for plants to grow.

These principles are summarised in tabular form in Tables 1 and 2 (adapted from the data of Huet and Varley). Although these may not be entirely comprehensive summaries, anglers will, I think, be able to put waters they know into one of the seven categories and find that characteristic features mentioned in the tables agree with those culled from their own experience. As far as rivers are concerned, it must be borne in mind that many have been modified by man, in particular in their lower reaches. The only waters difficult to fit into the scheme are navigated canals. Their particular characteristic is that they are disturbed by passing craft and plant life is prevented from becoming established. They are probably quite well oxygenated but very turbid. Reservoirs would come under the first two categories of still waters. Ponds are variants of the third type; they are usually small and occur in agricultural land, but it is difficult to make a clear-cut definition of 'a pond'.

We have thought about some of the features of different types of freshwater environments and have briefly mentioned the sorts of effects these are likely to have on the plants and animals, and of course the fish, living in them. Just as the physical arrangement and structure of a group of houses or a block of flats may affect the lives of the people who inhabit them, either individually or as a community, so the factors discussed in this chapter affect the lives of the inhabitants of lakes and rivers. Only those who are well adapted to the conditions under which they live can survive and flourish. Our freshwater fish illustrate this remarkably well – particular species being adapted to different sets of conditions. Anglers know well enough that what is good for coarse fish is not necessarily good for trout, and, within the coarse fish themselves, what is good for the carp is not necessarily so for chub and dace. After paying some attention to the question of food in Chapter 3 we shall think in more detail in Chapter 4 about some of the adaptations which fish species show to life under different conditions.

Chapter 3 – Food

FOOD WEBS

Whether one's taste is for sausage and beans, fish and chips, or something more elaborate, the uses to which any of these are put are very much the same. To recall what we said in Chapter 1, food is a living organism's source of building materials for growth and fuel for energy. The angler's primary concern is probably to consider how and on what his fish feed. In Chapter 1 some of the background relating to the need for food and its production was sketched in. In Chapter 2 physical conditions in lakes and rivers and some of their effects on residents were considered. Let us now give some attention to communities of freshwater plants and animals with which the angler is likely to be familiar and try to shed some light on the feeding relationships of all the members, and in particular the ways in which food reaches the fish. If the right sort of food fails to reach them there will be no prize specimens!

First, one cannot emphasise too much that it is the green plants of any community which are the mainstay of the whole food economy. It is they which are able to capture the sun's radiant energy and use it to manufacture sugars from carbon dioxide and water (photosynthesis). Then with the addition of nitrogen and phosphorus they are able to make proteins. All other living things (apart from a few specialised bacteria which are able to synthesise certain organic substances) are dependent on them either directly or indirectly.

The green plants are called, logically enough, the primary producers. The organisms which feed on the green plants are the consumers (or sometimes named secondary producers). Animals which directly eat green plants are called herbivores and those which eat other animals carnivores. A carnivore which confines itself to a menu of herbivores is termed a primary carnivore, one which eats primary carnivores a secondary carnivore and so on. There are also many animals which are very adaptable and may feed on both plant and animal material; these are the omnivores. If the latter consume mainly dead material we can call them detritus feeders. This particular group has been found to play a remarkably large part in freshwater communities. The relationships between all these different sorts of hungry individuals are expressed in a network or flow diagram form and can be termed a food web. The following diagram Figure 10 illustrates a simplified food web. The arrows, incidentally, always go from eaten to eater.

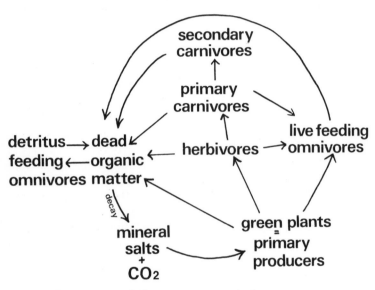

Fig 10 Simplified foodweb in a lake or pool, river or stream. (CO_2 = carbon dioxide gas.)

The Plants
Rooted plants can be thought of in three groups:
  (i) the submerged plants, *eg* canadian pondweed, water starwort
  (ii) the floating leaved plants, *eg* floating leaved pondweeds and water lilies
  (iii) the emergent marginal plants, *eg* reeds and bulrushes.
Very few herbivorous aquatic animals actually eat these plants directly and it might appear that they were rather wasted in the general economy of the freshwater community. They have, however, three important functions. One is that they provide surfaces on which algae grow, another is that they provide cover

40

for animals and, thirdly when they die down in autumn they contribute to the detritus on the bottom.

The most important primary producers from the herbivores' point of view are the algae. Covering stones on the bottom, or on the sides of canals, or sometimes floating in the open water grow the filamentous green algae, for example the blanket weed. Other algae consist simply of single cells or groups of cells floating near the surface. Amongst these are the diatoms and flagellates. They cannot be seen with the naked eye and live an entirely floating existence but sometimes are so numerous that they give a greenish pea-soup appearance to the water.

Not to be forgotten altogether are the terrestrial plants, trees and shrubs which live around the water's edge. These shed leaves into the water and therefore contribute to the detritus on the bottom (see dead leaves from overhanging trees).

The animals can be regarded as falling into three groups:

(i) those which live crawling about on the bottom or on the surfaces of the larger plants, *eg* leeches, worms, crustacea, insect larvae, snails, protozoans

(ii) those which swim actively in the open water, *eg* insects and, of course, fish

(iii) the microscopic floating ones, *eg* protozoans, crustacea, rotifers.

Of course there may be some of the bottom-living ones which occasionally take a swim, *eg* shrimps and dragonfly larvae; and some of the swimmers may spend some time resting on the bottom.

Bacteria are found everywhere, but in largest numbers on the bottom in the mud and detritus. They bring about decay.

Fungi are also found largely in the bottom detritus and cause decay.

Perhaps the least familiar of all these creatures are the minute floating plants and animals. These are known collectively as plankton, the plants being termed phytoplankton and the animals zooplankton. Most of the planktonic creatures live near the surface for it is here that there is light which the tiny floating plants need for photosynthesis.

Dead leaves from overhanging trees —it has long been suspected and recently shown experimentally by N K Kaushik and H B N Hynes (1968 and 1970) that dead leaves form an important source of food for many bottom-living creatures — particularly the invertebrates like the shrimps, water slaters, and insect larvae. They provide the best source of food when they have just begun to decay due to the activities of bacteria and fungi, and it has been shown that of the trees elm, alder, oak, beech and maple, elm sheds the best leaves as far as the invertebrates are concerned – they decay fastest and are definitely preferred by shrimps and water slaters (hog lice).

Drift fauna in rivers and streams – the groups we have considered above do not take account of the fact that in flowing waters many of the bottom-living creatures get swept out into the current and there may be a considerable drift of small creatures flowing downstream. The importance of this drift fauna as potential food for fish has been brought out recently by J Elliott who has been studying this during the past few years.

ANALYSIS OF GUT CONTENTS

A great deal can be found out about 'what eats what' by intuition and general observation of which animals nibble plant material and which chase insects and so on. A more objective and scientifically accurate picture can be obtained, however, by actually analysing the contents of the intestines of animals. In the case of fish it is a matter of cutting open the stomach, emptying it and examining the contents. The aid of a microscope is needed to identify small creatures, and the many fragments which are often all that is left of the larger food items. Expert knowledge of invertebrate anatomy is required to

---

Fig 11 This diagram illustrates the sort of food web operating in a productive pool. (See also Fig 10.) The large fish represents an omnivorous fish such as roach (though note that its food preferences may change according to its age, see Fig 12).
b Mineral nutrient salts entering pool in inflow stream. (Mineral salts also come from the process of decay on the bottom, not shown here but see Fig 10).
c Submerged plants.
d Floating leaved plants.
e Emergent marginal plants.
f Plankton (microscopic floating plants and animals, some of the latter being herbivores and some, *eg* rotifers, carnivores).
g Algae growing on stones.
h Bottom dwelling animals. Some of these will be herbivores (*eg* snails), some will be carnivores (*eg* dragonfly and beetle larvae, not shown here) and some detritus feeders (*eg* water slaters and shrimps).
i Dead organic matter from plankton and all other plants and animals.
j Dead leaves from overhanging trees.

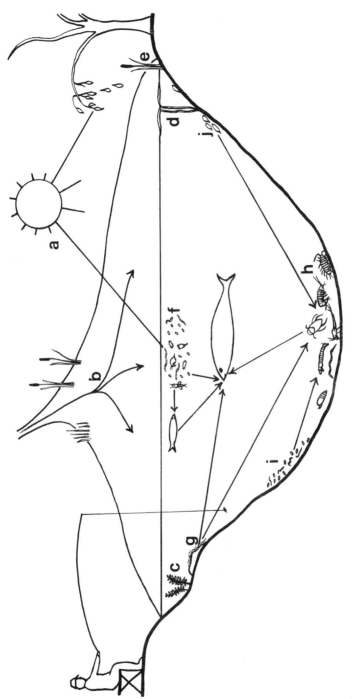

identify everything accurately. However hard a worker in this field tries not to miss things there are certain small soft bodied creatures which may well pass unnoticed through no one's fault. Examples are midge larvae, small annelid worms, leeches and flatworms. It is probably fair to say that the results of gut analyses are subject to a certain number of errors of this nature. But, bearing this in mind, the results of many peoples' studies have given us a great deal of information about who eats who in a typical lake or river. Figure 11 shows in outline how the plants and animals in a pool or lake fit into a food web. This basic pattern will of course vary under different conditions and according to which species of fish are present, but in a moderately productive stretch of water with a good variety of plants this is the type of food web which will exist.

FEEDING HABITS OF FISH

The focal point of all this as far as anglers are concerned is the fish. Looking at food webs shows us how the fish are dependent on a complex interrelationship between many species of plants and animals. But, as the angler is well aware, different species of fish have different feeding habits. It would be worthwhile then to have a closer look at the fish themselves. Many studies have been carried out on the feeding of our freshwater fishes. The subject is very fully discussed in 'British Freshwater Fishes' by M E Varley. Surveys such as Hartley's study of the River Cam fish (1948) and work in Poland reported by Stangenberg at the 2nd British Coarse Fish Conference have confirmed, what no doubt anglers have always known, that of our British coarse fish none can be considered true herbivores and only one species is a true carnivore – the pike. The rest are omnivorous, the cyprinids tending to favour plant matter rather more than the salmonids and the perch. This is perhaps to be expected when their respective habitats and modes of life are considered (see Chapter 2).

Of individual species three seem to have hit the headlines in terms of attention paid to them by the fish biologists. The trout, the perch and the roach have made frequent appearances in the scientific literature as subjects for detailed study. Species such as the tench are very much Cinderellas. This is no doubt explained by the trout's popularity as a game fish and by the fact that perch and roach are often problem fish in angling waters, attaining large numbers but stunted in individual growth. The facts that the scientific studies have thrown up,

44

however, are of great interest. One important point that emerges is that a fish's diet may change during its lifetime and may do so quite markedly and quite characteristically for a particular species at certain stages of the life history. This is particularly so for the roach and the perch. The pattern of feeding for roach is illustrated by Figure 12. One outcome of these feeding changes is

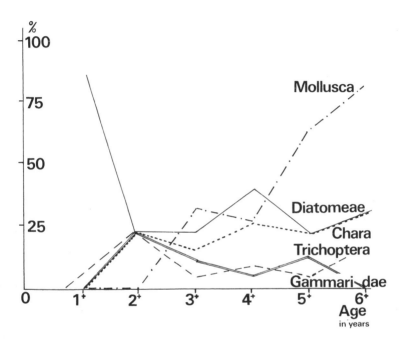

Fig 12 Graph showing how the food of the roach changes with its age. (From M Stangenberg in Proceedings of the Second Coarse Fish Conference, 1965.)
Mollusca – snails
Diatomeae – diatoms (= tiny plants mainly in plankton)
Chara – an alga growing rooted to the bottom
Trichoptera – caddis larvae
Gammaridae – shrimps.

that if a particular favourite item is missing when the fish need it most, *eg* molluscs at between two and three years (or at a length of about 15 cm as other studies have shown), their development may be seriously affected. A shortage of molluscs may well be the explanation for some cases of stunted growth in roach populations.

Another interesting feature of feeding habits revealed by research on predominantly carnivorous fish is that there are often many empty stomachs. This confirms the angler's painful knowledge that feeding is but sporadic! Carnivores tend to eat large meals and stop feeding while they are being digested while herbivores tend to be continuous croppers and munchers.

From the available literature the following attempt at summarising the feeding habits of our important species has been made (Table 3).

Table 3 List showing the main food items of our most common fish species (based on a number of gut analysis studies done in this country and on the Continent)

Pike	mainly other fish	a specialised carnivore
Trout	a wide range of small animals, *eg* small fish, insect larvae, crustaceans, molluscs	an unspecialised carnivore
Eel	a variety of small animals similar to the trout (but often not such a wide range)	an unspecialised carnivore
Perch	mainly insects and crustaceans when young, then small fish as perch grows larger	unspecialised carnivore changing with age
Roach	diatoms, filamentous algae and other plants and a variety of insects and crustacea when young; molluscs added when roach reach 15 cm in length	omnivore changing with age
Dace	diatoms, filamentous algae and other plants, insects, crustacea and molluscs	omnivore
Bream Tench Carp	diatoms, filamentous algae and other plants; insects, crustacea and molluscs – particularly those burrowing in the bottom mud	bottom-feeding omnivore

PRACTICAL ISSUES IN THE MANAGEMENT OF FISHING WATERS

Before leaving the subject of food the question 'How does one ensure that the fish get what they need?' must be answered. Another way of putting this is 'Does one corner of the food web channel available resources in an undesirable direction so that the best food bypasses the fish in which the angler is interested?' One link in a food web which might be undesirable in this way is that made up by small fish such as loach, gudgeon and stickleback. It is possible that they compete seriously with the fry of larger fish, consuming food which the angler would prefer to be going to the species he is after. It is on this sort of problem that scientific research can prove particularly valuable. Provided detailed and carefully worked out information can be obtained, guidelines for the management of angling waters can be put forward, *eg* in the case just mentioned it might prove useful to control (if possible) the numbers of small fish species if it were established that their presence was unfavourable to other species.

One general point brought out by a study of feeding habits is that variety is likely to mean greater stability for all. Since most of our fish are omnivorous the greater variety of species of plants and animals on which they can feed the better. If one type of invertebrate is off the menu for some reason, the fish can turn to something else until such time as the supply of the favoured item returns to normal. A survey by Witcomb (1963) has shown that different plant species tend to provide cover for different types of invertebrates. This survey needs to be followed up by more work along the same lines and in fact a similar kind of study is at present being carried out at Reading University. If Witcomb's conclusions are confirmed and found to apply widely then a beneficial course of action would be to make sure that one's angling water contains a good variety of the sort of plants that provide cover, even though they may not be important as actual items of food. This is particularly important in rivers which are subject to so-called improvement schemes. The clearing of banks and the cutting of aquatic plants may improve the river's function as a land drain but may ruin its potential as a rich fishery. Not only does this sort of clearance reduce the variety of habitats for food organisms, it also denudes the river of breeding sites for the coarse fish. Anglers must press for the retention of a good variety of water plants at least in some parts of our rivers.

BAITS

One of the most interesting things from the angler's point of view which is shown up by the scientific work on the natural feeding habits of fish is that most species are very adaptable and varied feeders. The fact that most coarse fish, and, indeed, the trout, will eat almost anything, and that they will even nibble at morsels with which they are not already familiar, is one of the secrets behind the excitement and interest of the sport – anything from caddis larva to Brand X luncheon meat might bring the luck today, and, of course, part of the art is making sure that the luck does come your way!

Perhaps the pike is the least variable feeder. He is a pretty exclusive and specialised carnivore and I should think that the angler would do well to stick to plugs, live fish bait or a metal lure for pike fishing. Trout are well known to fix their attention on any one source of food that happens to be particularly abundant, for example a large number of flies of a particular species in the process of hatching from the pupal to the imago (adult) stage. Thus the fly fisherman does well to keep an eye on what is hatching, or what species are swarming over the water before he chooses his fly. Coarse fish other than the pike are varied omnivores and on the whole feed on the bottom (with the exception of rudd and perch) but again if there is a particular source of food in abundant supply, even at the surface, they may make use of it. Richard Crudgington, a well-known Cheshire naturalist angler, has reported a case of bream being seen feeding on a shoal of fry, indeed having a fine old time chasing and devouring them on an early summer's evening. Perhaps then the use of a small metal spinner or small dead fish bait for coarse fish early in the season is not a bad idea.

Presentation of the bait is important, in particular for visual feeders. For example, the pike will be attracted by spotting something moving through the water in the natural way a small fish does – likewise for the bigger perch. In a river, casting upstream ensures that the bait will be seen drifting downstream as do many small creatures in a river. It should never appear to be tethered or dragging in the current.

Groundbaiting is obviously a good way of attracting fish to the swim where one wants them – but to overdo it is to lose the advantage of the fish having a small nibble and then following it up with a search for more of where that came from. Cloud groundbait is useful in tempting bottom-feeders up into more open water.

Beyond these comments it would be ridiculous to lay down any hard and fast rules. Anglers have their traditions, but it seems to me, as a non-practising observer, that the most successful fisherman is the one who is prepared to think for himself and be always trying new baits whether natural or artificial. Acquiring natural baits may mean some extra work – but with a small net made from a circle of wire and a piece of netting material sewn into a bag shape and attached to the wire (I made a perfectly good one from a pair of ladies' tights – but first make sure the lady of the house has finished with them!) One can fish up some of the bottom-living creatures to use as bait. As for artificial baits – be prepared to think of new ideas.

ENERGY FLOW IN FOOD WEBS AND IDEAS ABOUT ECOSYSTEMS

Give a few moments to pondering on life in a lake and one is soon brought face to face with the realisation that all its facets – the water, the mud, the plants and animals, the sunlight and shade, the inflow and outflow of streams – are all part and parcel of a network in which all the components are inter-dependent. This complete network is termed an ecosystem. The links between the parts of the system are made by virtue of a flow of energy and/or materials passing from one part to another. In recent years ecologists have been busy drawing attention to this and to the fact that the links between the

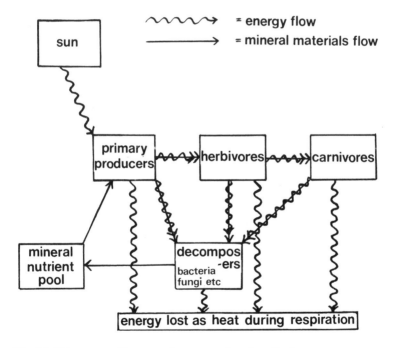

Fig 13 Diagram to illustrate the energy flow in a freshwater ecosystem. (Adapted from Edward J Kormondy, *Concepts of Ecology* © 1969) – by permission of Prentice Hall Inc. Englewood Cliffs, New Jersey, USA.

physical and chemical environment and the living organisms (or vice versa) are just as intimate and important as the links between the living organisms themselves. This sort of approach encourages us to think about freshwater ecology in terms of whole systems rather than just studying separately the parts which make up the whole. Ultimately the ecosystems ecologist would have us simulate things in a computer in order to work out what effects might be produced by altering inflows or outflows at certain points – just as a business management consultant might investigate the industrial system of a factory. A great deal more needs to be known in detail about the incredible complexity of the workings of life in a lake or river before the angler can call the computer to help him in his management problems. Perhaps the main importance of the ecosystem concept at present is simply to hold before us the reminder that everything is interlinked and that we cannot separate ourselves and the fish from either the other living things or the chemical and physical features of the environment. A diagram such as Figure 13 helps to bring this home.

THE EFFICIENCY OF NATURAL SYSTEMS AND THOUGHTS ABOUT HATCHERIES

In Figure 13 certain features of the flow of energy and materials are illustrated. One feature is that at every link there is considerable wastage of energy. The wastage is due to loss of energy as heat which is generated during the plant or animal's general activity. (Considerable amounts of heat are produced in respiration just as it is in the combustion process in a motor car engine.) A consequence of the loss at each link in a food web is that animals have to eat far more than they might appear to need for growth. Since by no means all of what is available to be eaten goes to making their own flesh, the total weight (biomass) of primary carnivores in an ecosystem is less than the total weight of herbivores. Likewise the total weight of herbivores is less than the total weight of green plants. In other words the biomass of organisms at any particular feeding level (trophic level) gets progressively smaller the farther away from the primary producers one gets. As far as the energy stored in this living material is concerned it has been estimated that plants store at most only 5% of the available radiant energy from the sun and that only about 10% of the total stored energy is passed on from one trophic level to the next. The herbivores assimilate about 10% (or 1/10) of the energy available in the plant food, and of this only 10% again is passed on to the carnivores, *ie* the carnivores get only 1/100th part of the energy stored in plants.

To return to talking about weights of flesh produced (which reflect directly the flow of energy from one level to the next) – in agricultural terms from 100 lb of grass in the field one can only expect to get 10 lb of cow flesh. In fishery terms, from 100 lb of phytoplankton one could get theoretically only 10 lb of zooplankton and from these creatures only 1 lb of fish fry using the zooplankton as a food source. If the fish fry were feeding also on the phytoplankton the weight of fish produced could of course be higher. Whether or not 'nature knows best', as the saying goes, it might well seem to us that the wastage is enormous. No less than 90% of the energy is lost at each level, though one must remember that a large proportion of this goes towards doing the work of keeping the organism active and alive. Nevertheless, the efficiency with which energy is passed through each link, to our minds, seems remarkably low.

On the other hand the actual growth efficiencies, *ie* the ratio of energy put towards growth to the energy actually consumed

$$\frac{\text{energy put to growth}}{\text{energy consumed}}$$

of individual animals may be higher than 1/10th. For example young broiler chickens and calves have growth efficiencies of about 35%. For every 100 lb of food actually consumed 35 lb go towards making flesh. The difference here is that the calculation is made on the food actually consumed and assimilated and of course under broiler conditions the food is supplied straight to the animal in the correct quantity and of the correct quality. What is more, the animals do not have to expend much energy in obtaining their food. The same principles apply in fish farms where growth efficiences may turn out to be higher than 10%. It seems to make sense to rear fish artificially if one wants the maximum possible yield. But, one must remember, as some ecologists fail to do, that the energy (and therefore the cost in financial terms) put into the making of the artificial food and the maintenance of the artificial environment have to be taken into account. In other words, if one wants high growth efficiencies, inputs of energy from elsewhere (most likely from our diminishing supplies of fossil fuels) may be necessary. Perhaps after all nature does know best. We cannot be sure until a great deal more research has been done.

GENERAL IDEAS ABOUT PRODUCTIVITY

The business of actually measuring the production of living tissue in natural systems has been a popular pursuit in recent years. Production, or productivity, is usually expressed in terms of weight of living tissue produced per unit area or volume per unit time, for example in grams/cubic metre/year. Sometimes it is expressed in terms of the energy producing potential of that material for example in kilocalories or Joules/cubic metre/year. Primary production is the production of plant material and since everything else is dependent on the plants the primary production usually reflects the production further up food chains. Thus by measuring the primary production of one's angling water and also possibly the surrounding land from which leaves could blow into the water one can get an idea of its general potential. But one cannot thereby always assume that if the primary production is high the productivity of individual fish species of interest from the angling point of view, will also be high. So many factors enter into the lives of fishes, and the resources of an ecosystem can so easily bypass one's favourite species and perhaps be used up by other species in which the angler has no interest. Suffice it to say, however, that if the primary productivity is high then the potential for coarse or game fishing is also high.

More will be said about this in Chapter 6. Meanwhile, a brief mention of ways of classifying bodies of water according to their productivity. Two words which are frequently used in this connection are eutrophic and oligotrophic. Eutrophic is used to refer to a highly productive body of water and oligotrophic to an unproductive one. They are very general terms and the dividing line between the two is a matter of opinion. It very often happens, however, that soft waters are oligotrophic and hard waters eutrophic, since, as we mentioned in Chapter 2, hard waters are richer in nutrient salts. On the whole the coarse fisherman is happiest with eutrophic conditions; it is here that most food for the fish is to be found. The game fisherman, however, does not want his waters to be too eutrophic – partly because highly productive waters invariably have muddy or silty bottoms. The organisms living there are less easily visible and salmonids, which are largely visual feeders, tend to be at a disadvantage. There are other reasons for salmonids doing better in oligotrophic waters as we shall see in the next chapter.

Chapter 4 – The Lives of Fishes

An experienced angler has only to look at a stretch of water and he will be able to predict what species of fish will be found there. What explanations can biologists offer for this? It appears that certain species are found only in certain places, and that each species is best adapted to living under particular conditions. It is fascinating to follow this up and try to sort out how it is that the trout flourishes in barren upland streams, and the carp in weedy lowland pools.

Referring to the tables in Chapter 2 we can remind ourselves that there is a grading in natural freshwaters between, on the one hand, conditions of strong current or wave action, low temperature and high oxygen content (and in general low productivity) and on the other hand conditions of gentle current, little wave action, muddy substratum, higher temperatures, copious plant life, lower oxygen concentrations (and higher productivity). There are certain fish species which are adapted to the extremes and certain species which are adapted to intermediate conditions.

Let us now follow this up in detail considering two groups of factors. First to be considered is basal metabolism – that is the basic level of general body activity. This can be related to the environmental factors of temperature and oxygen supply. Secondly breeding habits can be thought about in relation to substratum and vegetation.

ADAPTATIONS CONNECTED WITH RATES OF BODY ACTIVITIES

Of the specimens of the human race with whom you are acquainted there are probably some whom you regard as easy going fellows, in fact whom you might almost go so far as to call idle, while there are probably some you know who are always 'on the go' to the point of having a mania for getting things done. It is more than likely that your lackadaisical friends have a low basal metabolic rate (BMR) and that your super-active ones have a high BMR. This means that in the former normal bodily chemical reactions tick over fairly slowly, whereas in the latter fairly fast. It is perhaps not quite fair to make a comparison between individuals of the human species and different fish species, but the principle of the matter is there – namely that different organisms have different rates of everyday normal activity. In the case of fish there are species differences in basal metabolic rate. No doubt there are also individual differences within each species, but the difference between the species themselves is most striking. There are those such as the trout which tend to be constantly living it up and there are those such as the carp and tench which idle their hours away in the peace and quiet of still waters and thick vegetation.

Fry (1957) has summarised the data for the basal metabolism of five species of fish; rainbow trout, speckled trout (now called by some authorities brook char), chub, perch, and goldfish. At any one temperature the basal metabolism of the rainbow trout turned out to be the highest and that of the goldfish the lowest. The speckled trout had the second highest rate, then came chub and then perch. Unfortunately precise studies have not been made of all species, but one could suspect that the carp, a close relative of the goldfish, would fall into the lower range, other cyprinids in the lower and middle ranges, and the grayling and salmon in the higher range. This prediction can be made on the basis of a knowledge of the general habits of the different species; the salmonids are well known to be swift and active swimmers constantly 'on the go' whereas the carp and its relatives are notably sluggish creatures. Evidence for the correctness of the prediction however will come forward when we consider the effects of changes of temperature.

Temperature as an important environmental factor was mentioned in Chapter 2. There was set out the principle that an increase in temperature speeds up chemical reactions and a fall in temperature slows them down. This applies to the reactions going on inside a living organism just as it does to an elementary chemistry experiment. Fishes are amongst the so-called cold blooded animals and their body temperatures are therefore directly affected by the temperature of their surroundings. A rise in water temperature produces a rise in body temperature and a speeding up of body reactions, and vice versa. For every fish species there is an average temperature above which death occurs. Death at high temperatures is probably due to the speeding up of reactions in the cells to a rate at which the cells can no longer obtain the necessary food and oxygen to maintain the chemical activity. The temperature above which life cannot be sustained is called the ultimate upper lethal temperature and it differs for different species. A large number of studies have been carried out and, on the basis of them, Varley (1967) has suggested three categories.

(1) Fish with upper lethal limits below 28°C (82°F), *eg* freshwater salmonids (trout, grayling).

(2) Fish with upper lethal limits between 28°C (82°F) and 34°C (93°F), *eg* pike, perch, ruffe, roach, gudgeon and the migratory salmon.

(3) Fish with upper lethal limits above 34°C (93°F), *eg* carp, tench and probably rudd and bream.

This all fits in with the evidence that there are species differences in metabolic rate. From what was said earlier it might be expected that, under standard conditions, one could go on heating up a goldfish for longer than one could go on heating up a trout, before the point was reached where activity was too great for food and oxygen to get to the tissues fast enough. (This point is probably roughly the same for all organisms, *ie* there is a fixed upper limit to the pace at which living tissue can function). One would expect this for the reason that, as was said earlier, the carp has a very low rate of living at the start, whereas the trout has a considerably higher one. As the temperature increases their rates of activity rise roughly parallel to one another and since the carp's has further to go than the trout's before coming to the danger point the trout will succumb first, *ie* at a lower temperature than the carp.

Here is a simple analogy – there are two men walking up a slope, one 20 metres in front of the other. Both men can run at the same speed. A charging bull suddenly appears behind them and, so to speak, raises the temperature. At this the two men start to run up the slope – little knowing that at the top there is a hidden drop into a pond! The man who was ahead to begin with will fall into the pond first (albeit safe from the bull!). He is the salmonid, and his companion is the cyprinid who is exposed longer to the increasing temperature of the situation and reaches his fate later!

Acclimatization

Mention should be made here of the ability of fish to become acclimatised to new conditions within certain limits. In other words if a fish is kept at a slightly higher temperature than it is used to in its natural habitat it will become tolerant to a higher range of temperature in general, and its upper lethal temperature will be raised. Likewise if it is kept at a lower than normal temperature it will become acclimatised in such a way that its upper lethal temperature will be lowered. But there comes a point when no amount of acclimatisation can alter the ultimate upper and lower lethal temperatures.

Lower lethal temperatures and overwintering

If the body tissues of a living organism are cooled to 0°C (32°F) (or actually to slightly below, since the freezing point of water

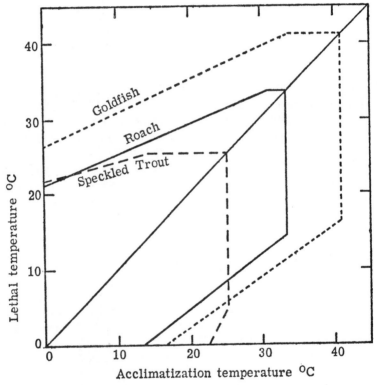

Fig 14 Temperature tolerance diagram for the speckled trout, roach and goldfish. The diagonal lines show the upper and lower lethal temperatures for fish acclimatised to different temperatures. The vertical lines at the right-hand side of the diagram show the ultimate upper lethal temperatures for each species. (Drawn from data from Fry, Hart and Walker (1946), Cocking (1959b) and Fry, Brett and Clawson (1942).) From M E Varley, British Freshwater Fishes.

containing dissolved salts is lower than 0°C) ice crystals form in the cells and cause irreversible damage and eventually death. The ultimate lower lethal temperature for most fish is therefore around 0°C. It is possible however for fish to survive at temperatures approaching 0°C in a state of hibernation, *ie* a sort of deep inactivity in which the cellular reactions are just ticking over at the slowest possible rate. Some of the coarse fish, *eg* tench are habitual overwintering species, often burying themselves in the mud at the bottom of a lake and remaining there without feeding all winter.

Figure 14 shows the results of experiments carried out to find out the lethal temperatures for three species under different conditions of acclimatisation.

Natural adaptations
One of the interesting things which emerge from all this is that our freshwater fishes are well adapted to their natural habitats as regards temperature tolerance. The waters of upland streams and rivers tend to be colder than those of the lowlands, and here in the cold swiftly running waters one finds the species which have high basal metabolic rates and are least tolerant to high temperatures but are very much at home in low ones. At the other extreme, waters of sluggish rivers and shallow lowland lakes tend to be the highest and here are found the fish with the lowest levels of activity and the highest upper lethal temperatures, *eg* carp, tench and bream. In between come those species which have an intermediate tolerance and they are found to inhabit waters which have a reasonable flow but never extremes of high or low temperatures, pike, perch, roach, being notable examples.

A brief thought here about heated power station effluents – if we apply what has just been said – carp, tench, rudd and bream stand the best chance of being happy in the lee of such an effluent.

Temperature and feeding
Since a fish's body temperature and rate of activity rises with an increase in the surrounding water temperature and falls with a decrease in temperature one might expect this to affect feeding habits – and indeed it does. A sudden rise in temperature may well rouse a population of fish into more vigorous feeding activity, particularly if it follows a rather cold period. Likewise a drop in temperature may make the fish go off the feed. This may

explain the fact that in many good fishing waters the fish tend to go off the feed towards the back-end of the year, that is around the end of October and beginning of November when winter temperatures begin to set in.

Oxygen requirements

The higher the metabolic rate the greater the amount of oxygen required. Thus one can see that our friend the trout is going to need more oxygen than the perch, who is going to need more than the goldfish or carp. (In fact the amount of oxygen used per unit time is often used as a measure of the metabolic rate). Again in this respect the turbulent well-oxygenated waters of mountain streams are ideal for the trout, but if transferred to a highly productive lowland lake which is subject to very low oxygenation at certain times the salmonids would be less happy. The carp, tench and bream on the other hand are better suited to conditions of low concentrations of oxygen on account of their general low level of activity. Transferred to a trout stream they would probably be happy as regards temperature and oxygen but being used to an idle sluggish sort of existence they would find the swift current a problem as also would be the lack of suitable breeding sites. This brings us to the question of shapes of fish and swimming habits and to the question of breeding.

SHAPES OF FISH

Water is something like 800 times denser than air. We mentioned in Chapter 2 that this is splendid as far as support is concerned – a fish does not need to expend a great deal of energy on holding itself up (indeed it also has the built-in inflatable bouyancy, the swim-bladder, to help with this). But when it comes to actually propelling itself through the water it is a different matter. The inertia of water is such that there is considerable resistance to objects being moved through it, and when the water is pushed aside it recoils in the form of turbulent eddies which buffet against the object and impede its movement even further.

Streamlined form is the answer to the problem and it is the streamlined body shape which has been selected during fish evolution. Fish are so successfully adapted in this way that experts in fluid dynamics can now look objectively at the form of fishes and say 'Yes – that's pretty good and it's the same form we'll use in our ship and aircraft design!'

Streamlined form reduces eddy currents and eases the fish's passage through the water. But the fish not only has to glide through the water, it has also to propel itself by sideways thrusts of the body. Putting these two requirements together, and coming to some compromise, one can see that the best shape is going to be fusiform (a sort of pointed cigar shape) but slightly flattened side to side, to give a greater lateral surface for the exertion of a thrust against the water. The more circular the cross section the better the streamlining but the less the fish can rely on body waves of movement and the more it has to rely on rapid movements of the tail fin for its propulsion. To take this further – the fish which is nearest to being circular in cross section will be a fast mover, a darter and rapid twister at high speeds, but there will be little flexibility of the body at low speeds since more or less all it can do is to thrust with its tail fin. The

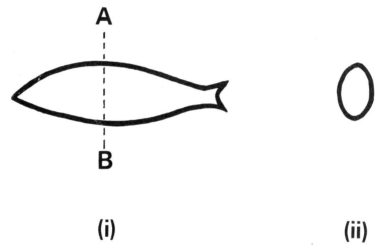

(i)   **(ii)**

Fig 15 Diagram to illustrate the 'compromise' fish shape.
(i) Longitudinal section showing torpedo shape.
(ii) Transverse section through A-B showing the side to side flattening which is not found in a torpedo.

deeper and more laterally flattened the body, however, the less important is the tail fin and the more important are sinuous movements of the whole body. Although less well-adapted to moving fast this fish can manoeuvre gently at low speeds.

A quick mental comparison of the shapes of the salmonids contrasted with the shapes of carp, tench and, in particular, bream will make us realise that this is in fact just the case in the adaptations of our freshwater fishes to their ways of life. The salmon and trout are torpedo-shaped when looked at from the side and are only slightly flattened from side to side in cross section. In contrast the bream is deep bodied and a good deal more flattened in cross section. The salmon and trout rely mainly on rapid and powerful tail movements while the bream relies on movement of the whole back part of the body.

Some experiments to substantiate these ideas have been done by Bainbridge (1963). He chose dace, goldfish and bream – the dace being rather more like the salmonids than other coarse fish. He designed the experiments to calculate the proportion of the total thrust exerted by the body and the tail fin respectively. In brief he found that 84% of the thrust was exerted by the tail fin

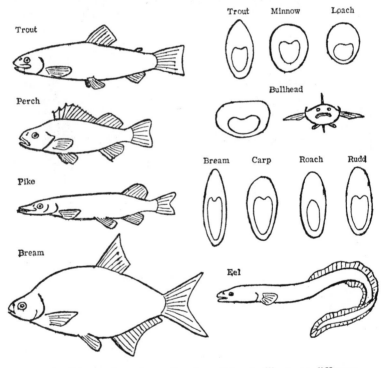

Fig 16 Outlines and sections of various fishes to illustrate different habits of life. The trout, minnow and loach live in rapid water and swim well; the bullhead lives on the bottom in rapid water; the perch is a moderate swimmer; the pike is a lurking predator which can make rapid darts, like an arrow; the bream, carp, roach and rudd live in slowly flowing or still waters; the eel is almost ubiquitous.

From M E Varley, British Freshwater Fishes.

of the dace, 65% by the tail fin of the goldfish and 45% by the tail fin of the bream. This is a nice illustration of the way in which fish are adapted in their shape and swimming habit to the fast hectic life of the rivers or the easy going existence of quiet weedy pools or slowly flowing waters.

BREEDING
Just as the evolution of a variety of species has involved the acquisition of adaptations to differing conditions of temperature and oxygen concentration, so also has it involved appropriate breeding adaptations. Which came first and was responsible for the initial separation into species occupying different habitats it is difficult to say. Probably all these features evolved together. Briefly one can say that the salmonids have developed a method of making a nest for eggs in the gravel which is found in plenty in their native haunts. The cyprinids on the other hand favour plants around or on which to lay their eggs. Even the size and hatching rate of the eggs bears a relation to the fish's habitat. Thus the salmonids lay a relatively small number (500– 4,500) of large eggs (3·5 or 7 mm in diameter) which take a long time to hatch and rely on the large amount of yolk in the egg. The young fry are large and able to feed on the same sort of food as the adults. The cyprinids on the other hand, as one might guess, do the opposite. They lay large numbers (up to 550,000) of eggs which are small (about 1 mm) and hatch rapidly into small fry which rely on the microscopic diatoms and crustacea which are in plentiful supply in a shallow lowland lake or in a sluggish river in summer. Again, the species we have come to associate with intermediate conditions of temperature and oxygen concentrations are also the species with breeding features intermediate between those described.

This chapter has, I hope, helped to show how any particular fish species can only flourish in a situation to which it is already adapted. Of course trout can live in a lowland lake and it is conceivable that carp could survive in a pool in a mountain stream. But they will not be at their best, for their natural leanings lie elsewhere. The trout may become acclimatised to higher temperatures but will be very sensitive to any decrease in oxygen – moreover it will only breed if there are inlet or outlet streams with patches of gravel. It has recently been established that barbel can live in still water pools and maintain good growth – but there is no evidence to suggest that it can breed there. The carp will likewise not be able to breed in the mountain stream pool without the establishment of dense vegetation. The discussion here has been brief, but these topics have been very fully dealt with by Varley and Chapters 2, 3 and 4 of 'British Freshwater Fishes' are recommended for further detail.

Chapter 5 – Fish Senses and Behaviour

The extraordinary antics and feats of 'cunning' which fish get up to are the stock-in-trade of many a good angling yarn. For thousands of years man has hunted fish, and the contest has seemed indeed to be a battle of wits between man and fish, with fish exercising a good proportion of the 'wits'. Observing the habits of some fish, man has also wanted to endow them with human moral qualities – Isaak Walton spoke of a 'loving and innocent fish', of 'lustful and chaste' fishes and of the salmon performing its natural 'duty' by hiding its eggs 'cunningly'. It may come as a disappointment to find that the scientific study of the brain and behaviour of fishes gives no support to the idea that fish can exercise cunning and wits in the sense of being able to think about a situation, and decide on a course of action by reasoning, as we are able to do. The behaviour of fishes is nothing like as complex as ours and can be explained in terms of sets of conditioned responses to particular situations. On the other hand, these responses are highly adapted to make the fish function successfully in its habitat, and the ways in which the fish is so adapted are marvellous and fascinating in their own right. From now on, we are not going to think of fish as possessing human qualities, but we are going to take them on their own terms, and try to get a fish's-eye view of what it is like to live without the powers of abstraction and conscious reasoning.

But here comes a warning to angler and scientist alike – though we may observe characteristic patterns of behaviour which we can explain easily enough, we may occasionally come across apparently extraordinary departures from the normal way of things. Let me illustrate this by quoting from an article by Richard Crudgington. After describing an extraordinary case of the surface of a lake becoming alive with bream performing the most unusual antics for several hours he says: 'An enthusiastic angler and naturalist of some years' experience . . . I have studied both aquatic flora and fauna of river and still water . . . yet in all that time even on this particular water I had never seen bream reacting in this way before'. Such is the bloody-mindedness of Nature (or, as the biologist would put it, the variability of biological systems). There are many question marks in the field of fish behaviour and the angler is in a good position here to turn scientist and try to put answers to them.

Before discussing behaviour let us consider the ways a fish receives information about its surroundings.

THE SENSES

As a fish goes about its everyday business of earning a living its main preoccupations are to find food and to avoid enemies. In fact, the fish's existence is mainly 'feeding in order to live in order to feed in order to. . . .' It must, in addition, of course, be constantly alert to potential danger and then there is the business of continuing the existence of the species – a mate must be found and reproduction accomplished. Indeed the fish's way of life might be summarised (in words to which no censor could object!) by what we might call 'the three f's' – Feeding, Fending-off foes, and Finding a friendly female.

To be successful in life the fish must be aware of what is going on round about and be able to react appropriately to a variety of situations. During the course of evolution animals have acquired ways of monitoring certain aspects of their environment. Thus, devices have been evolved for detecting light waves, sound waves, chemical substances, and changes of pressure, temperature and salinity. The fish is well equipped with such monitoring devices – these being its sense organs, eyes, ears, nasal sacs, lateral line system and a variety of sensitive nerve endings in the skin. These sense organs are connected up by the nerves with the fish's built-in analyser and computer, the central nervous system, consisting of brain and spinal cord. A nerve consists of a bundle of thread-like fibres, each fibre belonging to a single nerve cell. Some of the largest known fibres in the animal kingdom have a diameter of $\frac{1}{2}$ mm but most are very much finer. The brain and spinal cord are made up of masses of nerve cells each with one or more projecting fibres which connect up with other fibres to form an immensely complex network of interconnecting units.

When cells in a sense organ are stimulated by light waves, sound waves (or whatever it is adapted to monitor) nerve fibres leading from the sense organ are electrically activated and electrical signals are sent along them into the spinal cord or brain. Although the actual functioning is not quite the same as in our man-made machines, the nerve cells with their long projecting fibres can be compared with the circuitry of a computer and the junctions between them with the transistors or valves. It is in the central nervous system that the information received from the sense organs is sifted and interpreted and from which messages are sent out to the muscles to perform appropriate responses.

In general the senses of fishes are very much like our own – but with a different emphasis, and with two types of sensory apparatus which we do not possess – the lateral line and sensitivity to electrical fields (in certain marine and tropical fish).

Eyes – and how much of me can they see?

The eye of a fish works basically like a camera. It consists of a light-proof container (the eyeball) with a sensitive film at the back (the retina) and an aperture at the front (the pupil, which is a circular hole). Behind the aperture is a lens (in the fish a spherical transparent structure suspended in place by muscles and ligaments). Light rays from objects outside are focused on to the retina by the lens. From the images cast on the retina in this way, information is sent to the brain by the optic nerve. The brain then gives the fish the sensation of seeing. The system is very much the same as our visual system, but on account of the physical properties of water, and often its turbidity, it seems likely that fish do not have clear vision over long distances as we do. The world probably appears rather murky to them, though some are certainly capable of detecting a considerable amount of detail in objects a few feet away – at least trout are, as any fly fisherman knows.

From principles of the physics of light it is possible to work out that with its eyes directed to the surface a fish can see objects outside the water in a circular window – the bending of light rays at the water surface means that an inverted cone of light reaches the fish's eyes, and the window is the base of this cone. Outside this window the fish sees objects on the bottom reflected as in a mirror. The following diagrams illustrate this.

If the surface is disturbed the fish probably sees very little at all! The eyeballs are moveable in their sockets and can be directed so that the pupils can face towards the bottom. When

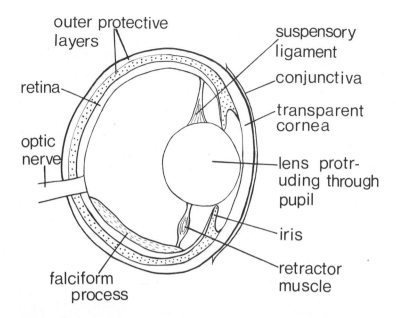

outer protective layers

suspensory ligament

retina

conjunctiva

transparent cornea

optic nerve

lens protruding through pupil

iris

falciform process

retractor muscle

Fig 17 Diagram of a vertical section through a typical teleost eye. (Simplified and redrawn from Walls, in Frost and Brown, The Trout, 1967.) Of the layers labelled protective layers, the outer is tough and fibrous, the inner contains dark pigment (helping to make the eyeball light-proof from behind) and many blood vessels (serving to supply the retina with oxygen). The suspensory ligament and retractor muscle hold the lens in position and the muscle helps to alter the position of the lens for focusing. The falciform process probably has a nutritive function like the layer behind the retina.

looking straight at the bottom like this the fish will receive on its retinae direct images of objects on the bottom.

An interesting suggestion about the possible effects of the trout's lack of eyelids has been made to me by Dr George Fearnley, an experienced Gloucestershire trout fisherman. He writes – 'Most trout fishermen would agree that on a bright, sunny day the chances of taking trout with a surface fly are negligible, except in shady parts of the river, *ie* under trees. On such days in clear spring-fed rivers, like the Gloucestershire Coln, trout can be seen busily nymphing amongst the weed, and only rarely taking the hatched *Emphemeridae*. In these conditions fishing a sub-surface nymph upstream is the only effective way of fishing. In the evening, when the sun is off the water, the surface fly comes into its own. In the past anglers

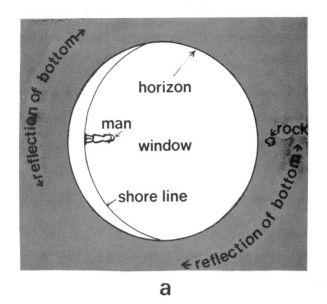

a

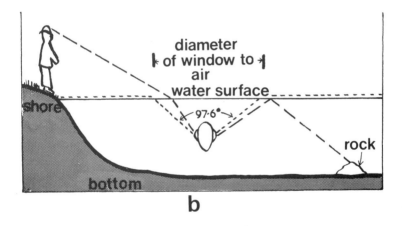

b

Fig 18 This illustration shows what a fish sees in the upward
direction when the water surface is perfectly calm.
(a) Water surface and aerial window as seen from beneath.
(b) Explanation of the window; rays striking the surface at an angle
within the window are refracted (bent) to the eyes of the fish but rays
striking outside the window from beneath are totally reflected. Within
an angle of 97.6° the fish sees out into the air; but outside this angle it
sees objects on the bottom reflected in a silvery surface. (Redrawn
from Walls in Frost and Brown, The Trout, 1967.)

have attributed their failure with the surface fly in these conditions to the ability of the trout to see the artificial fly for what it is and also to the visibility of the attached gut or nylon. More recently it has seemed possible to others and to myself that the real explanation is that trout are blinded at the water-air interface, or else that, since they have no eyelids, it is more comfortable to capture nymphs than to rise to the surface.' Is it possible to prove this one way or the other? Perhaps one could fit artificial eye-shades of some sort to a sample of experimental trout and see if they came to the surface under bright illumination more frequently than fish without the artificial eyelids. This would be a start towards trying to find out what is the true explanation.

There is some evidence that fish have colour vision of some kind. (See p. 74 for further details.) The experimental evidence suggests simply that there is the potential in the fish eye for distinguishing different wavelengths (*ie* colours) of light. This does not mean that we can say for certain that the fish sees the world exactly as we do in terms of colour, but simply that it seems likely that they can distinguish some differences in the colours of things.

The lateral line system – and the patter of fishermen's footsteps
This is a set of fine fluid-filled tubes set in the skin and opening to the outside at intervals by tiny pores. Within the tubes or canals are patches of sensory cells, each patch with a small particle of a

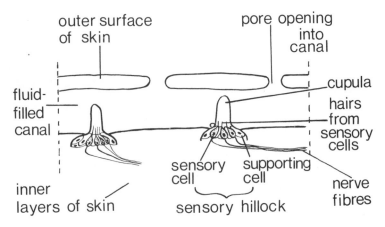

Fig 19 This drawing illustrates the structure of the lateral line system. The diagram represents part of the lateral line canal in longitudinal section (scales are not shown). (Drawn from a variety of sources.)

jelly-like substance (the cupula) attached to it by hairs projecting from the sensory cells. When the fluid in the tube moves the cupulae are made to swing about, the sensory cells beneath are affected, and electrical activity is set up in the nerve fibres leading from them. The cupulae are made to move by slight low frequency vibrations in the water. The system thus acts as a water disturbance detector or water pressure change detector. The sorts of disturbance detected are somewhere intermediate between sound waves and an actual contact of the body with something more solid than water, for example, the disturbances caused by the clomping of heavy footsteps!

Ears – and what about chatting to my neighbour and listening to the radio?

After I had spoken at great length at a meeting of anglers about fishes' hearing, and even about the structure of the ear, a voice piped up 'but where is a fish's ear?' A good question – where indeed!? So, the first point is that a fish's ear is simply a fluid-filled sac of gristle-like material lying in a capsule at the side of the skull – one on each side. It is called the 'inner ear'.

There is no outward sign of its existence. As in all vertebrates, one of its jobs is to detect vibrations passed from the water into the bones of the skull and thence into the fluid in the ear. From movements of the fluid it also monitors changes in position of

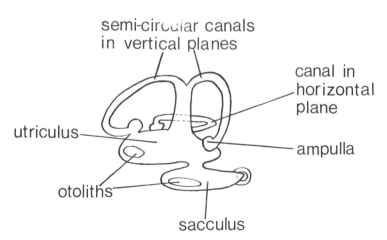

Fig 20 This shows the structure of the ear of a cyprinid fish. (Drawn from a variety of sources.)

the body, so that information can be supplied which can be interpreted and used (by the brain) to maintain balance.

In the part of the ear concerned with balance there are three semi-circular tubes (canals) curving out of the main part of the ear at right angles to each other. Each one has a small swelling, the ampulla, at its base containing a patch of sensory (sensitive) cells connected with nerve fibres going to the brain. In contact with the sensory cells lies a small solid particle rather like a cupula in the lateral line system. (In fact the inner ear is thought to have evolved from the lateral line system.) As the fish's head moves so the fluid in the canals moves and causes the cupulae to swing about. For movement in one plane the cupula of one of the three canals may be affected and will be made to swing to one side, stimulating certain of the sensory cells so that they send electrical signals along their nerve fibres to the brain. For movement in the opposite direction cells on the opposite side of the sensory patch will be stimulated; likewise for movement in another plane sensory cells in one of the other canals will be affected. Thus the brain receives patterns of information about the movements of the body.

In addition to these cupulae there are three larger particles each lying close to a patch of sensory cells within the inner ear. These are the otoliths and they have a function similar to that of the cupulae in the ampullae of the semi-circular canals. The otoliths are strengthened by calcium salts so that they appear bony. These are the 'ear-bones' used in age determination – not to be confused with ear ossicles which occur outside the inner ear.

Soundwaves, which are vibrations set up at a distance and transmitted through the air and water, probably act in the lower part of the inner ear, the sacculus, by stimulating groups of sensory cells. It is thought that different cells may be affected by different frequencies of sound.

In addition to this basic apparatus, some fishes have extra equipment. This is comprised of a set of ossicles or tiny bones, which help to amplify sounds and direct them to the ear proper. In carp these are particularly well developed and form a chain connecting the swim-bladder (which possibly acts as a sort of resonating box) with the inner ear capsule. Hearing in carp is reported to be particularly good and they are reputed to be able to detect sounds of between 60 and 6,000 cycles per second. So, when you're out for carp 'mum's the word' – and if you take your transistor radio I think it should be on only very softly since human speech and the music we listen to is made up of sounds of frequencies well within the 60–6,000 cycles per second range.

About the exact hearing capacities of other species we are as yet less certain. As with so many of the topics discussed in this book this is a field where there is plenty of room for further research!

Pressure change detectors in the skin (the tactile sense)

All parts of the skin are supplied with special sensory nerve fibres the ends of which are adapted so that they are stimulated by pressure changes at the skin surface. These are of obvious value in detecting previously unnoticed objects in the water. They are also useful in detecting currents, in particular the direction of flow of strong currents in streams and rivers. There is evidence that the trout uses the tactile nerve endings in the skin of the head to give it the information it needs to 'hold station' against a current.

Olfactory Organs (taste and smell organs) – and lacing the bait

A fish's nose goes by the elaborate biological name the olfacory sacs or nasal sacs. That these can be put to good use is the assumption made by those commercial concerns which produce patent fish attractants – substances smelling to us vaguely of strong kippers and to be applied to one's bait to make it more effective. Whether in fact these rather expensive aids to successful fishing are all that they are advertised to be is a matter for dispute. There is no doubt, however, that fish have a considerable capacity for detecting a variety of chemical substances in the water. Nicotine and engine oil appear to be among them – hence the angler's belief that it is unsatisfactory for a smoker to fix his bait immediately after smoking and unwise for someone who has just dealt with a boat's engine to handle the line. I think these beliefs are probably well-founded.

In most of our freshwater fishes the nasal sacs are small pits containing numerous folds of tissue. Each pit has an inlet and an outlet to the outside world. Water is made to flow through the pits by the action of numerous waving hairs or cilia. Different types of sensory cells lining the pits are probably stimulated by different substances. Here again more needs to be known in detail about how things work.

There are also chemically sensitive cells in the lining of the mouth and these are responsible for allowing the fish to obtain information about the chemical nature of objects actually taken into the mouth.

It seems likely that fish use their olfactory sense in everyday life a good deal more than we do. Current research on the way migratory fish find their way back to their home river indicates that smell plays an important part, and that these fish learn to associate certain chemicals with home waters, and can detect them on their way back.

Detection of changes in barometric pressure

The experienced fisherman knows that if the barometer is falling fishing conditions are going to change. From anglers' observations there is certainly evidence that fish can detect atmospheric pressure changes, but how they do it and what exactly are their reactions is open to question – a good subject for the angler/scientist to look into.

Detection of changes in temperature

This is possibly accomplished by nerve endings in the skin combined with changes in the activity and sensitivity of other organs. Again here is something which is by no means fully understood.

Pain

Many is the angler who must have wondered to what extent a fish feels pain as we do. One can, of course, never be certain about the sensations experienced by other animals simply because we cannot communicate with them about such things. (Even with the ability to communicate, there are certain difficulties about being certain that one's own experience is the same as that of another individual.) However, if we observe animals (fish included) we will notice that they react in very similar ways when strong stimuli such as sharp, strong pressure or abrasions, or electric shocks are applied – that is they will recoil, shudder and sometimes squeal and almost inevitably try to escape. This indicates that a sensation of some sort is being produced which makes the animal need to avoid the cause of the sensation. We can guess that this is an unpleasant feeling very much as we feel it. The validity of the guess is supported by other evidence that our nervous systems and sense organs work in very much the same way as those of our animal relatives. It seems fair to say then that fish almost certainly do feel pain.

Here, therefore, a plea to all anglers to exercise great care in the removal of hooks, and to be sure that hooks are not used which are larger than necessary. As we shall see from the next section, fish learn about being caught and the less pleasant it is the less readily will they take bait in future.

WAYS OF STUDYING FISH BEHAVIOUR

Before going on to discuss the ways in which the fish uses and responds to the information picked up by its sense organs, let us think briefly about what we could actually do to try and find out in a scientific way something about fish behaviour. First and foremost we should try wherever possible (not always possible!) to keep our fish in natural conditions. Who knows – mere captivity might have a distorting effect on an experiment.

The first steps are to present our experimental fish with a variety of situations or stimuli and note what happens. If an identical response is always given to a certain stimulus, at all ages from smallest fry to adult, the response is most probably a simple reflex, inherited and innate.

Further investigation will lead us to see if we can teach our fish to carry out this response when presented with another stimulus—in other words—can we make it learn, *ie* condition its reflexes? A possible experiment would be to attempt to train our pet goldfish to swim to a certain part of the tank for its food at the flash of a light, regardless of whether food was actually there or not. The technique is to present food at a certain spot, together with the light, repeatedly for several days and then see what happens when the light is flashed alone.

There are other ways of studying the senses and behaviour of animals, some of which require a steady stomach and no strong anti-vivisectionist views. In this connection, it should be emphasised that for these experiments the fish is always anaesthetised and the greatest care is taken to ensure that no undue pain is caused. There are two main types of experiment which can be done, one being to remove a certain part of a sense organ or the nervous system and see what happens without it. For example (a non-fishy one) if a cat is dropped 20 ft down a dark shaft it will land on its feet. But if its semi-circular canal system is removed by a small operation, it will no longer land on its feet. This provides evidence that it is indeed the semi-circular canals which enable the cat to keep its balance. There would not be much point in dropping a fish down a shaft! But I will leave it to the reader to think up experiments one could do with fish along the same lines.

The other type of experiment involves the use of sophisticated electrical apparatus – in particular, tiny micro-electrodes which can be inserted into the nervous system or sense organs and can actually record the electrical activity taking place in individual nerve fibres. This sort of technique has been used in the study of colour vision. For example, by testing different areas of the goldfish retina it has been shown that electrical activity is caused in certain patches of cells by light within one particular range of wavelength, *ie* of one particular colour, whereas other cells are stimulated only by light of other wavelengths. This indicates that the fish eye has the wherewithal for detecting different wavelengths of light, that is, has the capacity for colour vision.

FISH BEHAVIOUR – INHERITED AND LEARNED

An angler recently told me of the following observation he had made. Floating on the surface of a pool was a crust of bread, and circling round it a wary-looking bream. After a few moments the fish began to nose repeatedly at the bread until it began to break up. Once it was reduced to small pieces the fish then proceeded to make a meal of it. Here, our friend claimed, was an example of a fish apparently 'thinking things out' and adopting a highly intelligent course of action in order to obtain some tasty morsels. Now, the interesting thing is that this behaviour can be explained in fairly simple terms on the understanding that fish have a basic set of innate reflex responses plus the capacity for these to be conditioned. In other words (though I have to admit that I know nothing of the history of the particular fish in question) it is possible to explain the behaviour of that fish without resorting to the suggestion that it was reasoning about it in the sort of elaborate way that we might employ. From what follows I hope to show how.

Inherited Behaviour

(i) Simple reflexes

Endowed with the equipment to obtain information about its surroundings the fish is also endowed with certain innate responses to particular stimuli. We find, for instance that if we direct a current of water at a young trout it will swim into it; if we give our goldfish an electric shock it will recoil; if we present a fish fry with a particle of food it will nose and nibble at it. All these response are simple inherited reflex responses. Examples

74

of reflexes in ourselves are the raising of the leg when the patellar tendon is tapped, the contracting of the pupil when a bright light is shone into the eye.

(ii) Instinctive patterns of behaviour

Many animals show complex sets of reflexes in certain situations. This type of behaviour we can call instinct or instinctive behaviour, for example the courtship and nesting of birds, and, similarly, the nest building, courtship and care of the developing eggs carried out by sticklebacks. The preparing of the redd by salmonids and the subsequent tending of the eggs is another example. In the case of these chains of reactions there may have to be a certain set of stimuli – a certain combination of circumstances – to set the inherited behaviour pattern in motion. Once going, however, the chain of automatic actions is followed inevitably to the end, unless inhibited by the lack of a correct stimulus at some stage, or unless modified by learning, as we shall see in the next section.

Learned Behaviour

(i) Conditioning

In addition to these simple reflexes, however, we find that a fish can learn from experience. It may learn to associate another stimulus with the original stimulus which initiates a reflex action or a train of instinctive actions. Alternatively it may learn, by trial and error experience of reward, or the opposite, to adapt its response towards obtaining the maximum rewards and avoiding harm. Two examples will serve to illustrate this conditioned reflex behaviour. Just as the famous biologist Pavlov trained his dogs to salivate at the sound of a bell, so one can train a fish to make a feeding response to a stimulus quite unconnected with the food itself. What Pavlov did was to ring the bell every time he presented the dogs with food. The original reflex was to 'water at mouth' at the smell of food, but after repeated experience the food came to be associated with the bell, so that eventually the bell was enough to produce the response.

The innate reflex of nibbling at potential food in the young fish is probably largely a question of simply nibbling at anything which by visual and tactile cues is small and solid. In an aquarium a fish will swim towards and nibble at flakes or pellets of food presented to it. One can then introduce another stimulus, eg a gentle tap on the side of the aquarium every time food is presented. The fish will soon associate the tap with food until eventually it will swim towards the surface and make a preliminary feeding response to the stimulus of the tap alone. This is an example of a conditioned reflex.

A second example involves training a fish to make a simple 'choice' and to make it to its own advantage, as a result of trial and error experiences. The sort of experiment which might be carried out by fish biologists is to make it necessary for the fish after a period of fasting to swim into a compartment at one end of the tank to get their food – this, incidentally, would involve an initial conditioning to get the fish to associate swimming to one particular end of the tank with finding food there. At first there could be two compartments both of which always had food at the end. The situation could then be complicated by, say, the left-hand compartment always having the food, and the right-hand compartment delivering an electric shock if entered. Results of experiments of this sort have shown that the fish will learn to almost always swim only into the left-hand compartment and to scarcely ever even attempt to use the right-hand one.

In some recent experiments with rainbow trout carried out by J W Adren, P T Grant and C B Cowey (1973) the fish were presented with the possibility of operating a trigger. If they worked the trigger they got food and if not they went hungry. It was found that they soon learned to operate the trigger to get their food. Thus, by trial and error experience that working the trigger gave a reward, while leaving it alone meant hunger, the fish became conditioned to operate the trigger.

Fish can learn, by experiencing reward or by suffering a painful or unpleasant happening (or a combination of both), to do what is most to their advantage in the game of survival. What is more, there is evidence that learning can take place after only one or two trial experiences (see p. 78).

(ii) Memory

Having learnt certain things, fish are able to retain what they have learnt for some length of time. Certainly experiments with goldfish have shown retention of conditioned responses for periods of several weeks without the necessity of retraining. The recent experiments with rainbow trout, mentioned above, demonstrated retention of the conditioning to operate a feeding trigger for three months.

With the ability to pick up information about its surroundings, to learn about rewarding and harmful situations and to remember these things for a considerable length of time, the fish is well equipped to find food, avoid predators, breed and survive to a good old age. It cannot invent technologies and build civilisations, but it can manage very nicely with the simple capacities it has.

Now to return to our fish which seemed to be so 'intelligent' about the bread. Most species of fish will come to the surface to feed at some time or other, particularly in the summer when there are insects about. Supposing that the fish in question, while at the surface one day, chanced to knock against a piece of floating bread, causing a bit to break off (on heavily fished waters such discarded slices are not an infrequent sight). The small broken bit would appear as a potential piece of food and the fish would probably 'smell' it and bite at it. It would probably be easily swallowed and digested and the fish would now have the experience of having been rewarded by eating a fragment from a floating object with a certain characteristic appearance and odour. The next time such a floating object presented itself the fish might approach it and nose at it, in the process again breaking a fragment off, eating it and finding it indeed a tasty morsel. If this were repeated once or twice more the fish would by now have learnt to associate nosing at a large white floating object until it breaks up with a rewarding feeding experience. Our angler's observation can thus be explained in terms of a simple conditioning process.

CAN A FISH LEARN TO AVOID BEING CAUGHT?

By now it is sure to be crossing your mind that fish must learn about the unpleasant experience of being caught, and that just as a goldfish in an aquarium tank will learn to avoid an electric shock, so fish in a heavily fished pool will learn to avoid the hook. Experienced anglers talk about fish being 'educated' in heavily fished waters. Some experimental work has been done in Holland by J Beukema (1969) on just this subject – in this case on tagged carp in pools previously completely emptied and restocked for the purposes of the experiments. The trials were carried out with the aid of local anglers, the procedure being to fish with a variety of tackle and baits. On catching a fish details of the capture were recorded and the fish marked before returning it to the water, so that in the event of a second capture it would bear information about the previous capture. The results of these experiments have shown (a) that it became progressively more difficult to catch fish, (b) that numbers of carp captured a second time were significantly below what might be expected if catchability was unaltered by a first experience, (c) that high proportions of individuals were caught only once, (d) that the effect of decreased catchability was evident even a year after a previous capture. Other experiments done by Beukema have shown that pike can also learn hook avoidance though it turned out that they learnt more firmly when artificial bait was used than when natural bait was on the hook.

Clearly these experiments should be repeated, and, if possible, with other species. But from the results of this particular set of experiments it looks as though carp can indeed learn to avoid being caught even after only one hooking experience, and, what is more, can remember about it for an awfully long time.

It must be remembered, however, that there is probably considerable variation within any fish population between individuals who learn quickly and forget slowly (the 'brainy' ones) and vice versa (the 'dunces'). The existence of slow to learn ones probably explains reports of fish having obviously been caught several times in succession. Dr George Fearnley tells the following story: 'Some years ago I was standing on a footbridge spanning a branch of the Coln. The time was about 8 pm. Five yards upstream a trout of about half a pound was rising to spinners. I hooked it on a ginger quill at the third or fourth cast, netted it out, unhooked it and returned it to the stream. The trout returned to its lie and lay on the bottom for several minutes, after which, to my surprise, it started again to take spinners. I cast the same ginger quill to it again, and after about a dozen refusals I again hooked, netted, and returned the fish. By then it was too dark to continue fishing. I have no witnesses; but this was the only fish visible and I observed it closely from the time I first caught it to its second return. On four occasions I have been broken by a trout and caught apparently the same fish, with my fly and a length of gut or nylon in its mouth. There is of course no proof that these were the same trout, but the presence of fly and nylon indicate that someone, if not myself, had recently hooked and lost them.' No doubt others may have had the same experience. The fish which do get caught several times are probably the particularly slow to learn members of the population.

There is little doubt, however, that though some may be slow to learn many can learn to avoid being caught. What the cues are to which they become conditioned is of course not known and that is perhaps a line for future research to take. Meanwhile what can the angler do (with all that extra cerebral equipment which gives him the capacity for complex reasoning) to prevent his quarry avoiding his advances so efficiently? Experiments with goldfish have shown that certain drugs induce amnesia, *ie* have a forgetting effect. But the practicalities of administering such chemicals to one's angling stocks defeat one on the grounds of finance, ethics, side effects on the fish and other animals, to mention but a few problems! One obvious aid to helping one's fish to forget about being caught is to encourage the retention and efficient enforcing of closed seasons. A few weeks of respite from angling may not do a great deal of good but it is bound to do some good.

Of course if fish were not put back after being caught the problem would not arise. In certain situations there is a case, for other reasons (see Chapter 6) for not returning fish to the water – but it would probably be unwise to make this a general rule, since waters differ so much in their population structure, intensity of angling and so on. Angling waters are rather like people – they need to be treated as individuals!

The only thing the angler himself can do is to vary his baits, his tackle, his techniques as much as possible. If the fish are always presented with the same set of cues learning will be quickly effected – but if cues are always changing then it seems likely that fish will learn less easily. May I conclude with the words of a wise angler and naturalist who has learnt from his own experience – 'be prepared to change your ideas, think individually and try new tactics'.

CAN FISH COMMUNICATE?
A simple answer to this question is 'yes'. They do it, however, in a way quite different from and much simpler than our complex system of spoken and written language.

Communication amongst fish probably consists of the exchange of simple sets of signals which contain information about potential danger or about aggressive or amorous intentions. The sending out of the signals and the response given to them by the receiver are all part of inherited patterns of behaviour. When fish communicate they are behaving instinctively, triggered off by particular stimuli or a particular set of circumstances.

Evidence that fish use their sense of smell in communicating with each other comes from some experiments with minnows in which all the members of a shoal showed a strong fright and escape reaction when a damaged fish was placed in their midst. Following this up it was discovered that certain 'alarm' substances are released from damaged tissue and that other fish, when stimulated by these, immediately take fright and swim away. This is a way in which an injured fish can say 'I'm being attacked – swim for your lives!'

Communication as a prelude to mating has been studied extensively in sticklebacks. Here the swollen shape of the female's belly is the signal which tells the male that the time is ripe for eggs to be laid and for him to chase the female through the nest and fertilise them. This is an example of visual communication – a very simple sort of sign language.

Many fish can produce sounds – indeed fishermen (mainly of the marine variety) have been known to listen for the presence of fish. There is little space here to go into the fascinating details of this, but the following examples show some of the ways that sounds can be produced. Most teleost fishes possess opposing sets of teeth in the throat region and these can be rubbed against each other to produce a grating sound. A fish which is rather expert at this is one splendidly named the white grunt. Several species of catfish produce sounds by the movements of fins and some, for example the sea horse, can rub together adjacent bones in the body to produce some sort of coarse noise. The swim-bladder is also used, usually as a resonator, but sometimes actually as a drum – the trigger fish is reputed to beat its pectoral fins against the region of body wall covering the swim-bladder. In fish where the swim-bladder opens into the oesophagus gurgling noises can be made by the expulsion of air – a sort of burping, to put it less politely. What such sounds mean, if anything, to other fish is a wide open question for future research to answer. The examples given are of fish species which are not amongst our common freshwater fishes. It may well turn out though that some of our more familiar fish have a means of simple sound communication. The obvious group to start investigating would be the carp family since carp have a more highly developed auditory system than other species.

The ability to produce a fairly strong electric field is peculiar to fish and many fish (again not the ones familiar to us) have sensory systems for the detection of electrical activity in the water. Again, about the importance of these to the fish little is known but it is possible that they can use their electrical apparatus to signal danger or courtship intentions to one another.

We can conclude then that by virtue of their sense organs for picking up information, the ability to process that information in the nervous system and to make certain responses to certain situations, plus their capacity to learn from experience, fish are well able to cope with life in their aquatic environment. Though they appear to us to do remarkably intelligent things there is no need to think of explanations in human terms – fish have their own ways.

Chapter 6 – Growth Population Numbers and Fish Productivity

How often does the angler come to the fishery biologist with the problem of a water containing large numbers of 'stunted' fish and very few good sized specimens, despite the fact that to the best of his knowledge most of the fish are well advanced in years. Perch and roach are reported to be particularly prone to this condition. Recently two examples involving other species have been brought to my attention – one of a pool containing large numbers of small bream and the other large numbers of small rudd.

These are the sort of situations in which thorough scientific study of the ages of fish, patterns of growth, and estimation of population numbers is of direct interest to anglers. Anglers want to know what it is which causes stunted growth, and what are the factors that encourage the largest possible number of fish growing at maximum rates. Scientific methods are not by any means capable of instantly coming up with a complete picture of a situation, or of supplying all the answers. But an attempt to measure accurately and scientifically age, growth, numbers in a certain reach of water and so on, is likely to be better than guesswork. We can hope that as more studies of this kind are carried out, scientists will accumulate information which will be useful in guiding fishery management policies.

The following sections describe the methods most frequently employed by fishery biologists for studying these things, and there is no reason why some of them should not be employed by anglers themselves to help in building up stores of accurate information about fish populations.

AGE DETERMINATION

You may have experienced being in the unfortunate position of being able to catch plenty of small fish but never a big one. There are two possible explanations for the persistent and annoying appearance of small ones; either there are only young fish in your water, or there are older fish but they have never grown beyond a certain size. The first thing one must do then is to determine the age of these fish. The most convenient method, and the one in most frequent use for both coarse and game fish involves examining body scales. There are two other methods involving examination of the operculum (gill-cover) and the so-called ear bones which are the small calcareous nodules or otoliths inside the ear. The latter requires rather specialised apparatus and preparation of the bones, so only ordinary scale reading and examination of the operculum will be described here.

Scale reading

Scales grow during the course of a fish's lifetime and they do so by the formation of numerous rings of bony substance around a central point. Each new ring is laid down round the edge of the existing scale. In summer these rings are regular and well spaced out but in winter are much closer together, and rather irregular. Thus there is a visible difference between summer and winter bands of growth. Winter growth appears as a dark narrow band when the scale is held up to the light, and is sometimes called a 'check'. Counting the number of winter checks will give the age of the fish in years.

Scales should be taken as gently and carefully as possible from a live fish. A good plan is to take, with tweezers, two or three scales from each fish – say one from just behind the gill cover on each side and from further back. The scales should be placed in a small labelled envelope and pressed flat as soon as possible. To examine scales I have found that placing them between two pieces of stiff cellophane, perspex or glass, holding them up to a light and studying them with a good magnifying glass gives adequate conditions for counting the number of annual checks.

Examination of the opercular bone

This can be used in conjunction with scale reading if one has a dead fish. As the gill-cover bone grows, a different type of bone substance is laid down in summer from that produced in winter,

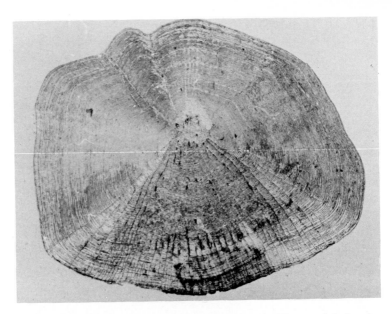

Fig 21 Photographic impression of a bream scale. The top right hand quarter shows the annual checks most clearly (as pale rings going round the scale). The author counts nine of these checks and estimates the fish to have been in its 10th year.

and here again, therefore, one finds summer and winter bands. If placed on a dark background and viewed with a light shining on to it the summer bands of bone appear light and the winter bands dark (the former are opaque and the latter translucent). The age is determined by counting the number of winter bands. Before examining the opercular bone it is necessary to rub off the skin and clean it carefully with a damp tissue or cloth. The bone can then be placed on a piece of dark paper and examined with good lighting from above and with the aid of a magnifying glass. If held up to the light the bands appear the opposite, ie summer opaque and slightly darker and winter transparent and bright.

Possible errors in scale reading

The number of yearly bands on a scale should be a true representation of the age of the fish. There are, however, certain snags to beware of. One of these is the difficulty posed by replacement scales. If a scale is knocked off during a fish's life-time, a new one may grow to replace it, and of course this replacement scale will only show growth rings for years subsequent to the accident. A guard against making a mistake on this score is to look at a number of scales from the same fish. Another thing which is important is to examine the scale carefully all round because sometimes there may be a slight irregularity of growth occurring in one region and obscuring the regular summer and winter pattern. Such an irregularity is sometimes termed a 'false check'. Something similarly termed a 'double check' may occur – this time all round the scale, and appearing as a band within which it is rather difficult to make out whether there are two years or one year's growth. Such an effect may be due to one poor year's growth. Here good magnification and lighting, and possible reference to other records of the history of the population, if available, are the only solution. Even bearing these snags in mind it is possible for experts to come to different conclusions when presented with a set of scales. It is really necessary for anyone doing scale reading to 'get his eye in' for any particular water.

Detailed criticism of the use of scale reading has recently been made by some Dutch workers who claim that certain fishes' scales are constructed in such a way as to produce interference effects and obscure the annual growth rings. The controversy over this is apparently unresolved – there is evidence to suggest that any such interference effects do not present a serious obstacle to coming to an accurate conclusion.

It would certainly be nice to find another quick, easy and more reliable method of age determination. Perhaps someone may hit upon some other way. Meanwhile conventional scale reading, provided one bears the possible errors in mind, is at present the best method we have for routine age determinations.

Length and Weight
In any study of a population designed to find out about the progress and patterns of growth, accurate measurements of the length and weight of individuals are required. Here a tape-measure and weighing scales are the tools of the trade, but it is well to remember that it is all too easy to be inaccurate. A steady hand and eye and plenty of patience are required! What is more, one must be sure to record whether one has measured to the base of the tail (which is the usual 'standard length'), the fork, or the tip of the tail fin.

POPULATION NUMBERS – METHODS OF ESTIMATING

To find out the number of fish in a particular pool or lake, or in a particular stretch of river or canal, is not a simple task. Ideally one should count every individual, but of course this is impracticable. One can get near to it when draining a pool and transferring fish elsewhere, but, where this is not the case, it is not a practical proposition to extract all the fish for census purposes. In order to get the most accurate estimate possible scientists have devised sampling methods, the results of which can be treated mathematically to calculate the total population figure. Two such methods are described here.

(1) The Lincoln Index method (mark – recapture method)

This involves taking a sample of the fish population (by netting or electro-fishing), marking a certain number of individuals and then releasing these marked fish back into the same reach of water. After some time has passed, during which it is assumed that the marked fish have dispersed freely throughout the water, another sample is taken. In the second sample there will be a few marked fish and the rest unmarked. Now, it can be mathematically established (and indeed practically demonstrated, if one experiments in small drainable pools or tanks), that the ratio of the number of marked individuals recaptured to the number of marked fish originally released is equal to the ratio of the total numbers captured in the second sample to the total numbers in the whole population. For the mathematically minded:

Where the number of marked fish released $= M$ (300)
the number of marked fish recaptured $= m$ (50)
the number of fish captured in the second sample $= t$ (250)
the total number of fish in the population $= T$

then $\dfrac{m}{M} = \dfrac{t}{T}$

$\therefore \ T = \dfrac{M \times t}{m}$

or in terms of actual figures $\ T = \dfrac{300 \times 250}{50} = 1500$

(2) Successive sampling or de Lury method

This involves taking several successive samples, this time not returning the fish to the water. As one would expect, the catches decrease each time, until, in theory, if one continued one would eventually catch nothing because all the fish had been removed. One doesn't in fact go on to the 'bitter end' because after a while it is possible to calculate from the data obtained (from say five or six successive samplings) what would be the result if one did go on to the bitter end.

For the mathematically minded again:

Where C = catch per unit effort

M = total population present before first fishing

S = sum of catches made in previous sampling sessions

K = coefficient of capture

then $C = K(M - S)$

If a graph of C against S is plotted, and the line extended to cross the S abscissa, then at this point where $C = 0$, $S = M$.

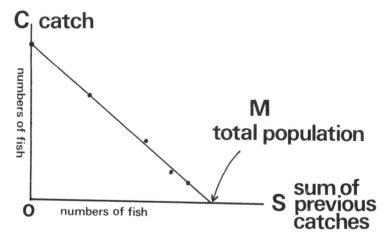

Errors in population estimations

As in most ecological methods, there are snags, and distortions may creep in. For example, in the Lincoln Index method the marked fish may for some reason not spread out evenly through the rest of the population, and this of course would seriously affect the accuracy of the final estimation. But keeping possible effects like this in mind, these methods provide a pretty reliable second best to actually counting every fish present.

Methods for sampling and marking fish are described in 'Methods for Assessment of Fish Production in Freshwaters' by W E Ricker, I B P Handbook, Blackwell Scientific Publications.

See following reference for a useful further discussion of the mark—recapture method.
Reference: D G Cross, 1972. The estimation of the size of freshwater fish populations. J. Inst. Fish. Mgmt. 3.

PRODUCTIVITY
(1) Primary productivity
Before talking about fish productivity, which is the quantity of fish flesh formed in a given reach of water during a certain time (usually now expressed in grams/cubic metre/year, though lb/acre are still resisting conversion to metric measurements), it is useful to know about the general productivity of the water, in other words the total amount of living material formed in the water expressed in weight per unit volume per year. To work out the total productivity of both plants and animals is a mammoth task. Since, however, the amount of animal production is directly related to the amount of plant production (assuming the absence of the introduction of organic matter from elsewhere) the measurement of plant production alone gives a good idea of the general productive potential of the water. The production of plant material is called the primary productivity and it can be fairly easily measured in still waters in a variety of ways.

For rooted plants in a lake or pool, the production can be measured by taking initial samples, weighing them and working out the standing crop say in grams/square metre. Sampling is then repeated after a period of time, and the productivity is equivalent to the increase in the standing crop over that time span.

For the planktonic plants, other methods are used which measure, in principle, the rate of photosynthetic activity. One method is to measure the rate of carbon uptake by the floating plants (mainly algae). To do this a source of radioactive carbon is made available to a sample of plankton. After a period of time the amount of radioactive carbon taken up into the algal tissues can be measured by special techniques. This gives an indication of how much they have photosynthesised and therefore how much new plant material they have manufactured. Another method is based on the principle that the more organic material being synthesised in the water, the more chlorophyll it will contain. Therefore a measure of the chlorophyll content of open water gives a measure of its primary productivity.

Very often when the primary productivity of a lake is being studied, only the floating plants of the open water are considered. You might say 'what about all those water lilies and reeds?'. Indeed, to make a complete primary productivity study one should include the rooted plants. But many rooted aquatic plants are not a direct source of food for the animals of the community (their main importance apparently is providing cover and also places for algae to grow). One can understand then why they are sometimes omitted from primary productivity studies.

(2) Fish Productivity

Having some idea of the potential total productivity of an angling water, one then would like to know what the actual fish productivity is. The sort of question you might ask is – 'are those fish we put in three years ago breeding and growing, and making good use of the available resources of the water?' For purposes of comparison it is useful to have a standard type of productivity measurement and, as described in the previous section, fish productivity is defined as the quantity of fish flesh formed in a given volume of water over a given period of time, now grams/cubic metre/year, though lb/acre is still in use. (Sometimes it is expressed not in weight, but in terms of the calorific value of fish body materials, ie the amount of energy released when the fish is totally burnt up. So one might see productivity expressed in kilocalories (K cal)/cubic metre/year) or in future as Joules/cubic metre/year. To measure fish productivity the history of a particular species population is usually followed from spawning time onwards. Samples are netted from measured reaches of water and individuals recorded for weight and age. The newly hatched fry in the population are then followed by sampling at regular intervals and a graph of the number of survivors at each sampling time plotted against their average weight. The curve so obtained is called the Allen Curve. The same thing can be done for every year class and, from the information these Allen Curves give, the total production of fish flesh over a given period can be calculated.

MORTALITY

The numbers of eggs laid and fry hatched is always greatly in excess of the eventual surviving fishable individuals. Large numbers of fish must die during the life history of any one year class. Gradual mortality shows up of course on an Allen Curve but very often the data are not detailed enough for the earliest stages of the life history when it is likely that the greater part of the mortality occurs. It would be useful to know more about these early stages in different species. What one would like to know is when and for what reasons does the main mortality occur and how does it affect productivity and future recruitment to the fishable stock. One problem here is that very small fry are exceedingly difficult to catch – they scatter rapidly as soon as they detect an approaching net. Ingenious work is being carried out on this problem at the Windermere laboratory of the Freshwater Biological Association by T Bagenal. He has devised a net which can be set on the bottom secured with a Polo mint. When the mint dissolves the net is released and floats gently upwards to the surface catching the small fry unawares! Using this new sampling device Bagenal hopes to get information about the early life of roach fry.

PRACTICAL APPLICATIONS

Having laboriously obtained all this information, how can it be put to good use in practical problems of fishery management? Let us think up some likely 'problem situations', and see how we could apply this sort of scientific data.

Stunted fish

First, let us return to the pool containing only small fish. Supposing that by sampling and determining the ages of the fish we had established that there was a normal age distribution, from first year fry to seven- or eight-year-old adults. We could then conclude that something was stopping growth at some stage in the life history. An interesting thing to do then would be to measure the primary productivity of the water, and compare it, if possible, with figures for the primary productivity of pools with similar numbers of fish all of which eventually grow to a good size. If the primary productivity turns out to be exceptionally low, then perhaps it is simply a question of lack of food in general which is slowing down growth. On the other hand, one might find that the general productivity of the water was high, in which case an explanation for the stunting would have to be sought elsewhere. It has been suggested that a shortage of a particular item in the diet at a time when food preferences change (for example the change to mollusc eating in roach) may be the crucial factor in causing stunted populations. Alternatively the channelling of food into other fish species which are of no interest to the angler might be a cause. It has even been suggested that over-production by rooted plants diverts raw materials (*ie* the nutrient salts) away from the phytoplankton and therefore away from the food chains on which fish depend. It may of course be that the cause of the situation lies simply in the effect of severe competition between the large numbers of fish for the available space and or food supply, that is, an effect of overcrowding. The answer in this case then is undoubtedly to reduce the fish population by taking out a good number of the stunted fish.

Stocking – numbers, productivity and size of individuals

Another question which might crop up at a club committee meeting is 'How many fish ought we to put into our pool in order to get the best fish productivity, with breeding and growth going on at optimal rates, and good proportions of large fish being produced?' To answer this question it would be necessary to have information about pools of the same size and primary productivity, but with varying numbers of fish in them. If one was fortunate enough to know of three pools of the same size and primary productivity which could be emptied and restocked with a single species, one could carry out an experiment. In one pool one would place a large number of fish, in another a small number and in the third an intermediate number. One could then compare their fish productivities over a certain period of time, maintaining of course a check on the population numbers in each pool. You would then have information which would be useful in making the decision as to what is the best population density for a pool of a certain size and a certain potential productivity. Experiments along these lines have been done in the United States by biologists Swingle and Smith. In one experiment they emptied three similar ponds and restocked them with varying numbers of bluegill bream. The numbers were 1,300, 3,200, and 6,500 fry/acre. One thing they found after about 6 months was that at the high population density growth was severely inhibited and there were large numbers of small fish. The growth of individuals was better at the intermediate population density and better still at the lowest density. The total fish productivity turned out to be about the same for all three ponds, namely around 300 lb/acre.

An experiment on a moorland fish pond in the English Lake District carried out by T T Macan has shown very clearly a similar effect. Initially there were fairly large numbers of small fish. The numbers were then reduced by netting, and, as the numbers of fish decreased, their average weight went up dramatically.

In other words, in overcrowded situations, although the standing crop at any one time may be high in terms of large numbers of small fish, growth of individuals can be at a standstill. This is not the situation anglers want – rather they want fish which are always growing into bigger and better specimens! They want a situation where elaboration of new fish flesh per unit time per individual fish is continually high. Precious little is yet known about the conditions and population densities which are best in these respects for our British species. Bagenal's experiments on early mortality in the roach are relevant here. They may help us to understand how the cutting down of numbers early on is essential for reasonable growth and development of individuals later. A start is being made but further studies need to be carried out in this country to help with making rational decisions on stocking.

There is one guideline, however, for those involved in managing fairly small lowland pools as coarse fisheries; take a hint from the results of the American experiments of Swingle and Smith, and Buck and Thoits, who have investigated various aspects of production of such fish as bluegill bream and carp in small pools. The results of their experiments have shown that whatever the initial numbers put in and whatever fertilisers were added to boost productivity about 500 lb/acre seemed to be the upper limit of fish production. In Buck and Thoit's experiments nine similar 1 acre ponds were each stocked with 1,000 carp fry, various fertilising programmes carried out and the fish production monitored over a period of several years. There was considerable variation in the figures obtained but after four years the maximum was 481 lb/acre and the minimum 321 lb/acre. As mentioned above, Swingle and Smith varied their stocking numbers and found that production levelled out at around 300 lb/acre. (Indeed 350 lb/acre is sometimes given as a stocking guideline by fishery management advisers in this country). Now, assuming that anglers would like, on the whole to be catching fish around 2 lb in weight or more and assuming an average fish production of 400 lb/acre for a fish such as carp, by doing some simple arithmetic one can work out that 200 such fish is all that a water can be expected to support. One cannot help suspecting from figures quoted by anglers and from discussing the matter with other biologists, that many waters are grossly overstocked.

A question relating to all this discussion about numbers is 'Should coarse fish always be put back?' The results of Swingle and Smith's and Macan's experiments mentioned beforehand suggest that there are occasions when it would be beneficial to have a period of not returning fish to the water in order to reduce the population density. On the other hand not all waters are overstocked. Ideally each should be treated individually and according to its needs. But before we know what the needs are we need facts to work on. Regular population estimates and records of what proportion of the estimated total population is being caught by anglers are important here.

Another question about which we could do with more scientific evidence is that of the value of introducing pike (if they are not there already!) to keep down the numbers of other coarse fish. Opinions are aired on this score but there has been precious little detailed scientific work done on the problem.

New stocks – from hatcheries or from other waters?

It is worth giving some serious thought to the question of where to look for new stocks. There are two things to take into consideration – first there is some evidence from work by a French biologist, Vibert, that the method of rearing trout fry under hatchery conditions has an effect on the fry in a variety of ways – the less natural the method (for example no gravel in the troughs) the less efficient were the fry in tolerating temperature extremes, holding station against current and avoiding predation. There is also evidence from Canadian work on salmonids that hatchery reared fish are at a disadvantage when introduced into a water where there is an already established population with which they have to compete. Secondly, I suggested that many waters are overstocked, so why not do another club a service by relieving them of some unwanted fish and at the same time buy them more cheaply than one might from a commercial source? This is not to say that fish farms are totally unnecessary but simply that they are not always the best answer to finding new fish for one's angling water.

Pool fertilisation

Something else a club might want to know is 'What can we do about water that is rather acid and low in nutrient salts?' The answer may be to go in for a period of treatment with limestone and fertilisers – obviously only practicable for pools and small lakes without much flow through them. Supposing a programme of fertilising was carried out, anglers would soon be able to tell roughly whether or not it had improved the fishing. It would be useful, however, to have a detailed record of the effect of fertilising, in terms of total fish numbers and fish productivity before and after. The information could then be used to guide others in making decisions about whether or not to use fertilisers. Experiments along these lines have in fact been done in the United States and also in Tasmania by Weatherly and Nicholls who studied a population of trout in Lake Dobson. In this Tasmanian study it was found that fertilising increased both population numbers and fish productivity. In other words it looked as though there was room for a considerable fertilising programme to improve the fishing.

In theory, treatment with limestone and phosphate/nitrate fertilisers might well help many of our upland lake trout fisheries, but, in practice, financial considerations have to be taken into account. The chemicals are expensive and have to be transported. In many cases it would undoubtedly turn out to be cheaper, and probably just as effective, to concentrate on manipulating numbers.

For small, accessible pools without much through flow (so that wastage is at a minimum) there may be a good case for fertilising if the water is acid and deficient in salts. Examples of such pools are those found in the border country between lowland agricultural regions and peaty moorland areas. It might be useful to include here some recipes.

Liming

Powdered limestone is a suitable form of lime to use. The technique is simply to go on putting it into the water until an alkaline reaction is obtained. Quicklime should be used only before fish are put into a water and then it should be applied two to three weeks before stocking. It is preferable to carry out a liming programme when the pool basin is drained, in which case lime is added to the mud until an alkaline reaction is obtained.

Fertilisation with nitrogen phosphorus and potassium

A recipe recommended by American workers (from Management of Lakes and Ponds by G Bennett) is as follows:
100 lb of standard agricultural N P K fertiliser per acre
10 lb of sodium nitrate per acre
These figures are for one application and it is recommended that between eight and fourteen applications should be made

94

between March and October. This is intensive treatment and obviously one would have to adjust to suit one's purse and the needs of the water.

Weed problems

In small, uniformly shallow highly productive lowland pools the excessive production of rooted plants (particularly canadian pondweed and curly-leaved pondweed) can become a severe problem. The chief difficulty seems to be that it makes fishing very hard because it tangles up the line. The best solution here is constant vigilance and regular yearly working parties to keep it under control by cutting it and dragging it out. It is most important to take the cut weed out because if it is left to fall to the bottom it will decay and on a warm summer's night might cause de-oxygenation (see Chapter 2). Err on the side of caution as far as chemical weed killers are concerned – the time will come when it may be safe to use certain herbicides in waters, but at the present time not enough is known about possible widespread side effects. What is more, the state of one's purse may well dictate the caution.

On the subject of clearing weed – you will have gathered from what has been said in Chapter 3, and in this Chapter, that, although rooted plants are not of direct importance as food for fish, they do provide cover for invertebrate food organisms, so that it is a mistake to be over-enthusiastic in your weed removal programme! I recently came across a water where the working parties had totally denuded the bottom of all forms of plant life. This is not to be recommended! What is to be recommended in the case of a shallow pool is to have it drained and deepened in a few places so that there are parts where light does not penetrate to the bottom so easily and the growth of rooted plants will be inhibited.

Another problem in highly productive pools is excessive growth of filamentous algae (blanket weed). Again, dragging it out with perforated buckets on lines is a possibility. Copper sulphate (bluestone) has been widely used to control algae and it is effective but must be used carefully. Twelve parts of copper sulphate per million of water is the maximum which should be used and the way of adding it to the water is to drag it in a sack behind a boat. The effectiveness of copper sulphate varies with the pH of the water – less is needed in acid water than in alkaline – and only trials can help one decide the correct dosage.

There are, of course, species of carp (not native to this

country) which feed almost exclusively and voraciously on weed, both rooted plants and floating algae. Experimental work is being carried out at present on the likely effects of introducing grass carp to our waters but again caution is the better part of valour and at the present time grass carp are not officially available for weed control.

In this chapter I have tried to show that by using scientific methods to try and find out as much as possible about the factors which govern the numbers and growth of fish in our angling waters we can come nearer to making effective decisions about stocking and providing good sport. If there appear to be more questions than definite answers it is because we are only beginning to get to grips with these things. The more research we can do in the future the better.

Chapter 7 – Maintaining Healthy Angling Waters

DISEASE

I have heard it said that anglers themselves suffer from an unmentionable complaint associated with sitting for long hours on cold, wet baskets. Be that as it may, fish, like human beings, are prone to ill health. It is difficult to define 'disease' precisely. Perhaps the nearest one can get to it is to say that disease is a departure from the best possible functioning of an individual. Of course it is difficult to say what is the 'best possible functioning'. One thing we know for certain is that when we ourselves are not in tip-top condition we have a feeling of being ill or 'off-colour'. Alas, we shall probably never know whether a fish has the same sort of feeling. We can, however, recognise sickness in a fish from certain symptoms, even though we cannot ask him how he feels.

SYMPTOMS OF DISEASE

The most obvious symptoms to look out for are as follows:

(i) Signs of irritation, *eg* rubbing against stones.
(ii) Abnormality of eye movements. If the fish is turned on its side, and the eyes follow this movement so that the pupil is visible when the side of the fish is viewed from above, this indicates that something is wrong. In a normal fish the eye will retain its original position and the pupil should be invisible from above.
(iii) Increased breathing rate and pale gills.
(iv) Discoloration of the body, spots or ulcers on the skin, and extra slime secretion.

WHAT IS DISEASE AND WHAT CAUSES IT?

Looking objectively at the many human diseases which have been discovered by medical science in the last century or two it seems possible to classify them into two groups.

(i) Diseases which are a disturbance of normal metabolism caused by an inherited defect in the system, or by the upsetting of the system due to accidental damage or to some sort of abuse such as drinking or smoking too much (these are human examples of course!).

(ii) Diseases which are caused by another living organism invading or infecting the body and becoming a parasite there. (A parasite is an organism which earns an easy living from another, the host. The parasite may live either inside or on the surface of the host and feed on substances in the host's body.) Not all parasites cause disease. In fact a well-adjusted parasite causes as little harm as possible for it is in its own interests, so to speak, to keep its host alive. Many parasites can, however, cause disease, particularly when they invade the host in large numbers. Examples of disease-causing parasites are:

Viruses
Bacteria
Fungi
Protozoa
Larger invertebrates, *eg* worms, flukes, lice etc.

DETECTIVE WORK ON DISEASES AND DETAILS OF VARIOUS PARASITIC DISEASES

Very little is known about the first group of diseases in fish but quite a lot is known about the second group. Perhaps the easiest diseases to study and come to grips with are the parasitic diseases caused by fairly large creatures such as flukes and tapeworms. Let us look therefore at one common disease of coarse fish and follow the stages by which biologists have discovered its course and its causes.

A good example to take is Eye Fluke (sometimes called Worm Cataract). The main symptom of this disease is a whitening of the eye lens. The obvious first step in the detective work therefore is to examine the diseased eyes microscopically. The cause of the damage is immediately apparent – hundreds of tiny flat oval-shaped worms living in the lens and sometimes also in the surrounding eye tissue. Such infestation of the lens must obviously mean impaired focusing capacity – the condition is indeed very like human cataract.

To any Sherlock Holmes worth his salt the next question is 'How do the worms get there?' Now if infected fish are kept with uninfected fish in an aquarium, the infection will not be passed on to the healthy ones. This should strike our detective as strange, since parasites need to find a new host before the current one dies on them. From this observation one can infer that once in the fish these particular flukes tend not to return to the water or to go in search of new hosts in which to live. One can also infer that either the worms are breeding in the diseased fish but that their eggs cannot be passed on to other fish without some intermediate agency or that the worms never attain adulthood but are in fact larvae which cannot themselves develop further and breed in the fish's body.

Information is now brought in from another source. While studying a parasitic fluke found in water birds biologists noticed that when these worms bred and produced eggs, the eggs hatched into larvae very like the ones found in the diseased eyes of the fish we have just been talking about. Here was the vital clue, and this was supplemented by the observation that if an

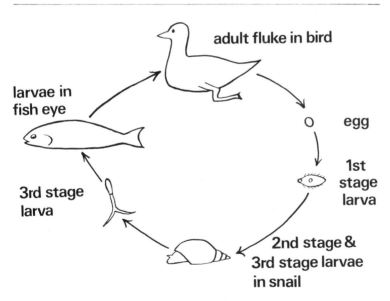

Fig 22 The life cycle of the eye fluke *(Diplostomum spathaceum)* is revealed in this diagram.

99

infected fish was eaten by such a water bird the worms from the fish established themselves in the body of the bird and became adult there. A further clue was provided by the discovery that freshwater snails often contained fluke larvae and that if healthy fish were kept with such snails the fish became infected. It was also found that the larvae left the snail, lived free in the water for a while and then entered the fish by actually burrowing into the skin.

The story can now be pieced together. A bird eats a fish infected with eye fluke; the larvae attain maturity and breed in the body of the bird; eggs are passed out into the water where they hatch into first stage larvae; these larvae enter a snail where they pass through two more stages of development; third stage larvae are then released into the water and enter a fish, by burrowing into the skin, and eventually become established in the lens of the eye ... and so the cycle goes on.

Many coarse fish which do not rely greatly on sight when feeding can tolerate a considerable number of fluke larvae in the lens without it greatly affecting their well-being. Infection is more serious for visual feeders such as perch, pike and the salmonids.

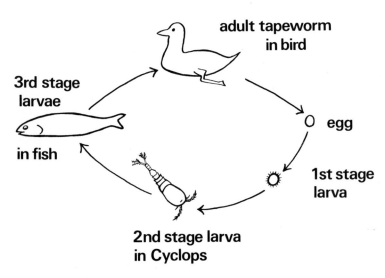

Fig 23 The life cycle of the fish tapeworm (*Ligula intestinalis*) is shown here.

100

Besides these worm cataract flukes there is a species of fluke which is the carrier of Black Spot Disease. This also has a bird and a snail as intermediate hosts in the life cycle.

Another common parasitic invader of fish is the tapeworm (*Ligula*). Its life history has been worked out in a similar way and is shown in Figure 23.

It remains now to say something briefly about some of the other parasitic diseases.

Roundworms (Nematode worms)
These worms are round in cross section, unsegmented and pointed at both ends (unlike the tapeworms and flukes which are flattened in cross section). Fish very often have an internal infection of roundworms without suffering any ill effects and only heavy infestation leads to illness.

Spiny-headed worms (Acanthocephalans)
As their name implies they cling to their fish host by rows of spines on the head. There are intermediate hosts in the life cycle as there are in those of flukes and tapeworms.

Crustaceans
The fish louse *(Argulus)* and the anchor worm *(Lernaea)*, an extraordinarily modified crustacean, both live as ectoparasites attached to outer layers of the body – the skin and gills. Again they only cause trouble if present in large numbers.

Protozoa
A disease caused by a protozoan infection is White Spot (blessed with the splendid scientific name of *Ichthyophthiriasis*. It is due to a tiny one-celled animal *(Ichthyophthirius)* covered with fine hairs (cilia). It penetrates the outer layer of the skin and lives in a small cyst in the skin tissue. It is this cyst which shows as the white spot.

Fungi
There are many fungi which infect fish – many of them also living on dead organic matter in the water. They consist of masses of fine threads (hyphae) which form a network inside the host organism or material. They reproduce and spread by means of spores and these spores are much more likely to become lodged in fish tissues where there is an injury to the skin.

The danger of fungus infection is therefore greatest after damage to the skin or in regions such as the mouth where the lining is thin.

Bacterial and Virus Diseases

Much research has been done on these diseases in fish but our knowledge is still abysmally incomplete. Many of them are rather similar to bacterial and virus diseases in ourselves in the sense that symptoms are much the same – for example red spots on the skin as in measles and chicken-pox. A healthy fish in excellent condition will normally have a good resistance to bacterial and viral infection because, as in ourselves, there are antibodies (substances in the body specially designed to fight off invading micro-organisms). The fish becomes ill only if the invading microbes become firmly established in large numbers.

The following examples are some of the most common and fairly well understood bacterial and viral diseases.

Columnaris – the symptoms of this are lesions or sores on the skin and gills. It is thought to be caused by a bacterium which also lives freely in the water.

Tail and Fin Rot – the tail and fins take on a rather ragged appearance (not to be confused with the results of a bite by another fish). This is also thought to be caused by a free-living bacterium.

Cotton Wool Disease or Mouth Fungus – results in a whitish swollen condition of the mouth and lips. It is not caused by a fungus but by a bacterium – though it may well frequently be confused with a fungus infection.

Furunculosis – this is a disease which commonly attacks trout and is prevalent in trout farms. It also attacks some coarse fish. The symptoms are characteristic skin lesions and internal haemorrhages. Its accurate diagnosis is only possible by making bacterial cultures from the sores. Indeed this is true of many of these diseases in which the symptoms are often very similar.

Fish tuberculosis – a similar condition to tuberculosis in man, though of parts of the body other than the lungs. It is caused by a variety of species of a particular type of bacterium known as *Mycobacterium*.

Pox Disease of Carp – this manifests itself by the formation of large spots on the skin. It is thought to be caused by a virus.

IDC (Infectious Dropsy of Carp) – the advanced symptoms of this are a generalised swelling of the body. The disease is now

102

considered to be a complex consisting of at least two virus diseases (i) spring viraemia of carp with inflammation of the intestine and general swellings and haemorrhages and (ii) carp erythrodermatitis, which involves ulceration of the skin.

UDN (Ulcerative Dermal Necrosis)—this is a disease thought to be primarily of salmon and trout. A recent epidemic started in Ireland in about 1965–66 but such epidemics of the disease have probably occurred before, for example the salmon disease reported in the Solway in 1877. The main symptom is ulceration of the skin and is thought to be initially a virus infection with a fungus infection following in its wake. It seems to affect fish as they are entering rivers and it also seems to be associated with low temperatures. As with many of these diseases, more research needs to be done before the picture is entirely clear.

CONTROL MEASURES

Anglers are clearly anxious that cures for these diseases can be discovered. The difficulties are obvious – for while human beings can all be contacted and asked to come forward for inoculations in the case of an epidemic, and while humans will go to the doctor of their own accord for treatment, by no stretch of the imagination could we get our fish friends to co-operate in this way even supposing we knew the cures for their complaints.

There are, however, some simple steps which can be taken in the right direction towards at least controlling the incidence of disease. They fall roughly into three groups:

(i) Chemical therapy – Most bacterial diseases respond to the same sort of drugs used in human medicine and there are various chemicals which are known to kill off invertebrate parasites of other types. In fish hatcheries these can be applied in solution in the water or in medicated feed. Obviously there are problems in applying these methods to natural waters and it would be dangerous to start adding drugs to angling waters in the hopes of curing diseases. Antibiotic injections have, however, been used to good effect on carp and perhaps there is a future in netting and injecting individual fish – though expensive and laborious!

(ii) Biological control – This can be used in the case of a parasitic disease involving an intermediate host. The theory behind this is to remove or control the numbers of the intermediate hosts so that the life cycle of the parasite cannot be completed and the disease is prevented from spreading and will eventually die out. In practice there may be difficulties here on account of the conflict of interest between bird watchers and anglers. Perhaps the fairest thing is for anglers to tolerate and indeed assist as much as possible in protecting birds on sites of special ornithological interest while bird watchers must be tolerant of the needs of anglers if something needs to be done about an epidemic of fluke in a water which is of less importance to them from the ornithological point of view. Control of snails is easier – the only problem there is that they are an important dietary item for some fish.

(iii) **Increasing the resistance of the fish stocks** – here exactly the same principles apply as do in 'keeping fit' oneself. All important are a good and varied diet, clean and unpolluted surroundings and the prevention of overcrowding. If fish have to live in dirty deoxygenated water along with far too many others they probably suffer from the same sort of 'under the weather' state of affairs as one does oneself after several hours in a traffic jam or cooped up in a stuffy overcrowded room at a boring party. One is much more likely to succumb to infection when one is under the weather or run down than when one is feeling on top of the world. Thus, for the same reasons, it is important that fish should have the best possible conditions to live in to keep them in good form and resistant to infection.

(iv) **Careful handling** – Infection by bacteria, viruses and fungi will take place much more easily if the skin is already damaged. It is most important, therefore, that if the fish are going to be put back into the water they should be handled as carefully as possible. Damage can so easily be done by hooks which are too large, hooks which are removed carelessly, clumsy landing techniques (for example dragging fish up the bank), and overcrowding in keep-nets. A sign that more notice is being taken of this sort of problem is the recent development of knotless keep-nets, designed to do as little damage to the fish as possible when they are jostling about inside. Another welcome improvement would be more widespread instruction for beginners in how to choose tackle, land and handle fish.

Reporting diseased fish

The best procedure for anyone with a disease of fish problem is to telephone the nearest River Management Division of their Regional Water Authority and report the incident. They will then be able to answer the relevant questions in order to establish whether it is necessary for the fish to be examined in the laboratory. If it is thought desirable, then arrangements will be made for the fish to be collected by a member of the Division's staff and it is then only necessary for the informant to keep the fish in a cool place (not a deep freeze) until collected. A River Management Division is unlikely to be very interested in an isolated fish found suffering from a disease. Instances should only be reported when a reasonable proportion of the population appear to be infected.

POLLUTION

So much has been said in recent years about this problem and the threats to natural ecosystems, that to add yet more words to those already uttered might well seem tedious. The word pollution, however, arouses strong emotions. It does so quite rightly. If ever there was a case for righteous anger it is over the fouling of clear waters, the mass destruction of wild-life and the spoiling of a sport. But, where emotions run high, attempts to ferret out the facts may be lost in a hurricane of hot air. It is useful to sit back sometimes and take a cool look, first at the different kinds of pollution and what they are actually doing, secondly at simple ways in which pollution can be detected and monitored by any angler, and, thirdly, what can be done to rectify a bad situation.

Before saying anything further – what do we mean by the word pollution? Let us define it and take it to mean the contamination of the environment with substances harmful to living things.

Pollution is not by any means a new problem. It has been the angler's bane ever since people started to congregate in large industrial towns, and convenient dumping places for factory wastes and human sewage had to be found. The Salmon and Freshwater Fisheries Act of 1923 came about largely as the result of pollution problems. It is to our credit that we are now far more concerned about the need for care in waste disposal than were our ancestors of 50 to 100 years ago, and great strides have been made to improve matters in recent years. It may not now be long before landing salmon on London bridge is a reality.

DIFFERENT KINDS OF POLLUTION

It is convenient to divide different types of pollution into three groups: (i) general organic pollution, (ii) toxic chemical pollution, (iii) thermal pollution.

Organic pollution is usually the result of an inflow of inadequately treated human sewage, liquid produced in the making of silage, or the sludge of excrement from intensive farming units, or a food processing factory. This organic pollution is dangerous simply because it may cause deoxygenation of the water. The offending effluent is full of organic substances which are the staple diet of microscopic organisms, chiefly protozoans and bacteria. What happens is that these creatures begin to feed on the organic matter and flourish and multiply in their millions. As living organisms they need to respire and of course they will use up oxygen. Large numbers of micro-organisms consume considerable quantities of oxygen, and thus where organic effluent is present there is a danger of oxygen supplies becoming depleted to a dangerously low level. Organic pollution spells death for fish by asphyxiation.

The state of organic pollution can be tested for by measuring the amount of oxygen taken up during a certain time by a sample of water. The bacteria in the water feed on the organic matter and use up oxygen. The amount of oxygen used up over a given period of time (often five days) is called the BOD (Biochemical Oxygen Demand). For severely polluted water this will be high (of the order of 300 for untreated sewage) and vice versa.

There is one type of organic pollution which is not necessarily connected with inflows of sewage, *etc*. This is the pollution produced by over-productivity (sometimes called over-eutrophication – see Chapter 3 for explanation of eutrophic). Some lowland waters can become so productive of rooted plants and algae that when, towards the end of the summer, the dead remains of these plants accumulate and bacteria and fungi get down to their work of decomposing, in the same way as for an organic effluent, de-oxygenation can be the result. A water becomes more productive when there is an abundant supply of nutrient salts such as nitrates and phosphates. Little detailed research on the run-off of farm fertilisers has been done and opinions here vary. It seems likely, however, that the use of fertilisers in large quantities just before or during wet weather may result in the chemicals being washed into rivers and lakes in larger amounts than usual. There they have exactly the same effect as they would on the fields and productivity is given a considerable boost. (See Chapter 6 for suggestions for controlling excess algal growth with copper sulphate.)

Attention should perhaps be drawn here to the rather slack

use of the word eutrophication. Eutrophication simply means becoming more productive. It has often been used by itself in recent years to imply a state of over-productivity giving rise to organic pollution. If this is to be conveyed then strictly one should use the term over-eutrophication.

Toxic chemical pollution

At one time anglers were to be seen 'eskimo fashion' poking holes down which to drop a line through great masses of detergent foam. Detergents are amongst many substances detected in freshwaters which must for the most part come from industrial processes, and which can be harmful to living systems. Such chemicals include the following: ammonia, cyanide, arsenic, zinc, copper, lead, chromium, nickel, cadmium, phenols, and a variety of pesticides and herbicides. These substances may be taken in and absorbed by fish and their effect is simply to interfere with the normal chemical reactions going on in the cells. Such interference may just mean general ill-health or immediate death. In either case it is the result of the effect of the chemical on the metabolism of the body.

There is also what might be called mechanical mineral pollution – that is the release of clouds of fine solid particles of, for example, china clay, coal dust *etc*. This state of affairs may affect fish directly, possibly by clogging up gills or indirectly by covering the bottom and destroying food supplies. Detergents were at one time a problem because they were not broken down in sewage plants (indeed they were sometimes responsible for killing off the sewage bacteria). Once released into a river they had the undesirable effect of preventing re-oxygenation of the water at the surface. This problem has now been solved by developing so-called soft detergents which are broken down by bacterial action during the treatment of sewage. Unfortunately an additional difficulty is that when broken down they release phosphates, and, as mentioned just before, extra phosphates may mean over-eutrophication. Research is going on at present into the development of adequate low phosphate detergents.

Thermal pollution

Those giant cooling towers are in fact not quite the ogres they look. Their job is simply to cool gases or steam by using a counterflow of cold water. They are not involved in any sinister brewing of poisonous potions! The water which passes through them, however, inevitably comes out at a higher temperature than that at which it was sent in (sometimes as much as 10°C (18°F) higher than normal river temperatures). This water then has to go somewhere.

Ideally the heat should be re-cycled but as often as not the warm water is discharged directly into a river. The results of this on the plants and animals at the receiving end have been studied by CEGB teams at Drakelow and Ironbridge and by a few other workers in this country, for example a study by D Cragg Hine on the effects of a heated effluent entering the River Nene at Peterborough. So far no disastrous effects have been detected. Very often the heated water stays near the surface and the fish themselves can take avoiding action. As far as the fauna in general is concerned, no changes have been observed in the balance of species which can be directly related to changes in temperature. There is some evidence concerning the effects on the life histories of certain invertebrates, for example the water slater *(Asellus aquaticus)* and a species of freshwater mussel have had their breeding activities advanced by at least one month below Earley power station. But otherwise there seem to be very few marked effects on the fauna.

So far as fish are concerned, increases in the rate of growth of fry have been demonstrated (see Chapter 1 for comments on recent studies of this) but the effect does not seem to be carried on into later life.

In the USA temperatures are higher and there is some evidence that this boosts productivity in general. Fishing in the canal leading from the Connecticut Yankee Atomic Power station has boomed since heated water was channelled into it. Twice as many fish are now caught there in winter as are caught in the main part of the river in summer! It has been found that if salmon were present they have tended to be replaced by coarse fish – which is not surprising in view of the ranges of temperature tolerance and oxygen requirements of salmonids. Potential harm seems, therefore, to be to game fish.

HOW CAN ANGLERS DETECT POLLUTION?

The fisherman soon knows about serious pollution – the awful and all too familiar sight of large numbers of dead fish afloat. But the important question is how to detect small amounts of pollution and forestall anything worse. I was once taken by a club officer to a pool lying downstream of an intensive pig-farming unit. Several months previously there had been a gross

Plate 1 Organisms useful in indicating the presence and severity of organic pollution.

1 Mayfly nymph (× 4). [N.B. Three 'tails' at the end of the abdomen. A stonefly nymph is similar in appearance but has only two 'tails').

2 Caddis larva and case (× 4).

3 Freshwater shrimp (Gammarus) (× 6).

4 Water slater (Asellus) (× 6).

5 Chironomid larva (× 6).

Table 4 Classification of Biological Samples (Biotic Index)

			Total number of groups present				
			0–1	2–5	6–10	11–15	16+
					Biotic Index		
Clean ↑ (As degree of pollution increases organisms in order of tendency to disappear) ↓ Polluted	Plecoptera nymphs present	More than one species	—	7	8	9	10
		One species only	—	6	7	8	9
	Ephemeroptera nymphs present	More than one species*	—	6	7	8	9
		One species only*	—	5	6	7	8
	Trichoptera larvae present	More than one species**	—	5	6	7	8
		One species only**	4	4	5	6	7
	Gammarus present	All above species absent	3	4	5	6	7
	Asellus present	All above species absent	2	3	4	5	6
	Tubificid worms and/or Red Chironomid larvae present	All above species absent	1	2	3	4	—
	All above types absent	Some organisms such as Eristalis tenax not requiring dissolved oxygen may be present	0	1	2	—	—

* Baetis rhodani excluded.
** Baetis rhodani (Ephemeroptera) is counted in this section for the purpose of classification.

110

Groups

The term 'Group' here denotes the limit of identification which can be reached without resorting to lengthy techniques. Thus the Groups are as follows:

Each known species of Platyhelminthes (flatworms)

Annelida (worms) excluding genus Nais   Genus Nais (worms)

Each known species of Hirudinae (leeches)

Each known species of Mollusca (snails)

Each known species of Crustacea (hog-louse [water slater] shrimps)

Each known species of Plecoptera (stone fly)

Each known genus of Ephemeroptera (mayfly) excluding Baetis rhodani

Baetis rhodani (mayfly)

Each family of Trichoptera (caddis fly)

Each species of Neuroptera larvae (alder fly)

Family Chironomidae (midge larvae) except Chironomus Ch. thummi

Chironomus Ch. thummi (bloodworms)

Family Simulidae (black fly larvae)

Each known species of other fly larvae

Each known species of Coleoptera (beetles and beetle larvae)

Each known species of Hydracarina (water mites)

Each known species of Hemiptera (water bugs)

Reproduced by permission of the Severn-Trent Water Authority from the booklet The Work of the Pollution Control and Fisheries Department, *Trent River Authority 1972.*

leakage of organic effluent into the inlet stream of the pool and the result had been the death of virtually the whole fish population. The farmer was duly penitent but who could guarantee that it would not happen again? Here was a case for every member of that angling club to help in keeping a wary eye open for any more leakage.

The biological study of organically polluted waters has revealed some useful principles. Certain creatures are more resistant than others to oxygen lack, so that as pollution gets worse so certain species disappear in an orderly sequence. By finding out which of these indicator species are present one can get some idea of the state of the water. The animals to look out for are the following ones: *Plecoptera* (stone fly nymphs), *Ephemeroptera* (mayfly nymphs), *Trichoptera* (caddis larvae), *Gammarus* (shrimps), *Asellus* (water slater), tubifex worms and red chironomid larvae (bloodworms). The Trent River Authority (now the Severn Trent Water Authority) and the Clyde Purification Board have collected many samples of these creatures in different waters, and in order to simplify the presentation of their results the Trent River Authority workers have devised a coding system called the Biotic Index. This works on a scale of 1–10, 1 representing highly polluted water and 10 representing clean water. Table 4 shows how this coding is applied to standard samples. You will notice that the first creatures to disappear in organically polluted water are the stone fly nymphs, mayfly nymphs and caddis larvae. The next to succumb are the shrimps, and the more resistant ones, the water slater and particularly the tubificid worms and chironomids, can tolerate quite low oxygen concentrations. Plate(s) 1 illustrates the indicator organisms.

The intention in talking about this is simply to explain what is meant by the Biotic Index. So often one hears such terms bandied about and one has not the slightest idea what they stand for. As far as the angler himself is concerned all that is necessary to do about detecting pollution is to dip a pond net (mesh size not less than eight threads per cm) into the water and run it along the bottom once or twice in several directions. As much mud as possible should be washed away and the contents should then be emptied into a dish of clear water and the living creatures carefully extracted (an old spoon is a useful piece of equipment for this) and examined. Books useful for identifying animals are listed at the end of Chapter 1. If stone fly and mayfly nymphs are present the water should be in good condition. If nothing is there but bloodworms one can conclude that the opposite is the case. Reference to Table 4 will give you some idea of what conclusions you can come to if the situation is somewhere in between. Another type of index sometimes used is the Diversity Index. For a detailed discussion of the advantages of using one or the other see reference at end of this section.

Detecting toxic chemicals is not such a simple task. If toxic chemicals are suspected it is best to send a water sample to your Regional Water Authority where there should be the equipment for assessing what is present.

Reference: P M Nuttall and J B Purves, 1974. Numerical indices applied to the results of a survey of the macroinvertebrate fauna of the Tamar catchment (southwest England). Freshwater Biology **4**, No. 3.

WHAT IS BEING DONE ABOUT POLLUTION?

It is encouraging that a considerable amount is being done now to study and control pollution in our rivers. The Department of the Environment has a Water Pollution Research Laboratory and the regional authorities are all bound by acts of parliament to see that pollution is controlled and that the interests of fishermen are protected and served. Some anglers may well feel that they do not get value for rod licence duty from their particular authority but admirable strides have been made in the past few years to come to grips with cleaning up our rivers.

Half the battle is to be able to detect the presence of harmful substances and to be able to predict accurately what they are likely to do to fish. With such facts at hand action can then be taken against the polluting offender.

In the last few years research has been going on apace into the choice of indicator organisms for organic pollution and on toxicity testing of fish for predicting danger levels of chemicals in our rivers. The development of a Biotic Index to assess organic pollution has already been mentioned. As far as toxic chemicals are concerned a good deal is now known about the lethal concentrations of certain substances. Most of the experiments done so far have actually been on rainbow trout chiefly because they are easy to keep and rear. As it happens they are probably a satisfactory choice as the few comparisons with coarse fish which have been made show that rainbow trout

113

are considerably more sensitive to most chemicals than their cyprinid relatives.

The index used for the assessment of the toxicity of chemicals is what is called the 48 hour LC 50 (the 48 hour lethal concentration). This means the concentration of the substance which kills half the experimental population in 48 hours. Although quite a lot is known about these values for individual chemicals under different conditions of oxygen saturation and temperature, there are certain outstanding snags. One important difficulty is that certain combinations of substances have different toxicities from those of the chemicals on their own. It has been found in general that combinations have an additive effect, so that concentration a of a substance x and concentration b of substance y are fractionally as toxic when by themselves as when mixed together, that is, in a mixture the 48 hour LC 50 is much lower than it would be for either substance alone. Another stumbling block is that although intensive surveys have been carried out in the last five years, it is still not possible to say what the long term effects of exposure to low doses of poisons are likely to be. A further snag is the appearance of entirely new synthetic chemicals – here new techniques of analysis have to be developed but this may not get done until considerable damage has been suffered.

There is of course the alarming possibility that there are chemicals in our waters for which tests are never made. This leads to the telling of the fairly well worn story about the anglers who have complained that their fish are being prevented from breeding because they have been unwittingly consuming contraceptive hormones. The question of how these are supposed to have got into the water will not be pursued here. But let it be a cautionary tale reminding us of the possibility of unforeseen dangers. The only answer is a watchful eye all round.

WATER ABSTRACTION

This is a truly serious hazard to anglers, in particular to game fishermen. Certain regions, notably Hampshire, have been harried by it in recent years. The particular danger is to trout streams, where, if water flow is reduced, silting up of the gravel beds favoured for spawning may occur, and pool haunts may become too shallow to support large numbers of fish. In lowland rivers, what were once good roach waters may become fit only for bream.

Water flow may be reduced not only by direct abstraction

from rivers but also by reducing the amount of water reaching the rivers as a result of bore-hole abstraction or the construction of reservoirs in the upper parts of river basins. An interesting thought was recently brought to mind by Lionel Sweet in a television programme. He has fished the Usk for 50 years and he pointed out that in the old days fishing was good for three to four weeks after a spate, whereas now it is a matter of three to four days. He attributes this to the large plantations of conifers on the hillsides of the valley. These probably take up a great deal of the rainfall which used to find its way into the river.

Certain tactics can be employed to increase the flow in trout streams by making banks of stones and boulders so that the water is directed to the centre of the stream. A short article has been written about this by R V Righyni (see reference at the end of this section). It would seem, however, that all efforts should be made to put pressure on the regional authorities to abstract water from as near the mouths of rivers as possible. Here the inhibiting factor is probably the expense of purification but surely that can be set against the lowered cost of piping, since a water source low down the river system is likely to be nearer to the industrial centres which it serves.

Finally, a reminder that every owner of water or land adjoining a river or stream has a right under common law to have the water come to him in its natural state both in quality and quantity. Wherever it is possible to take legal action (and here the Anglers' Co-operative Association has done good work in connection with pollution) it should be done.

Reference: R V Righyni, 1971. Improving trout streams after water abstraction. Salmon and Trout Magazine **193**, pp. 212–215.

CONSERVATION – THE ANGLER AND THE COUNTRYSIDE

There has, in recent years, been an upsurge of interest in the countryside and in the importance of conserving our natural heritage of wild plants and animals whose habitats are constantly threatened by the activities of industrial society. If we were to allow the variety of our natural flora and fauna to diminish we should be the worse off in more ways than one. For one thing there is scientific evidence to suggest that variety helps to give stability to natural systems. The interplay of large numbers of different species helps to damp down disastrous fluctuations in population numbers and helps to prevent plagues of pest organisms which may affect farmers' crops. Secondly, there are many people who derive great pleasure and benefit from watching and studying wild plants and animals – and I am sure that this includes many anglers who enjoy observing what is going on around them during the many hours of quiet by the water.

Unfortunately conservationists do not always have friendly inclinations towards anglers because they feel that the activities of fishermen and the management practices they sometimes employ are not always in the interests of the conservation of other species, particularly birds. I see no reason why this should be so. As I have explained in this book (especially in the chapter on food) the presence of a good variety of plant species in an angling water is of great importance. Although anglers may well want to clear away some of the water plants to provide swims, to avoid diverting too many of the available nutrients into rooted plant growth which is not a direct source of food for fish, and also to prevent the danger of deoxygenation when the plants decay, it is not in the interests of a healthy angling water to denude the place of its variety of plant life. This applies both to submerged plants and also to fringing vegetation which is of course dear to the bird-watcher as it provides nesting sites. The angler will not want too many shrubs and trees at the water's-edge, partly because of the difficulty of casting and partly because over-shading may inhibit good growth of water plants. On the other hand there is no reason why trees and shrubs should not be left alone in some places, particularly on the north side of the water. As was mentioned in Chapter 3 leaves falling from surrounding trees are in fact an additional source of food for aquatic animals. Perhaps bird-watchers could concede that a few spaces be cleared and overhanging branches lopped to make casting easier. As far as the kinds of plants in and around the water's-edge are concerned a good variety is of benefit to the food webs in the water from the angler's point of view and to the variety of habitats for birds from the ornithologist's point of view.

We mentioned earlier in this Chapter the problem of water birds being the carriers of certain parasites of fish. To any bird-watcher who is reading this book, therefore, a reminder that large numbers of certain water birds may be a nuisance to anglers, and it is only fair that they should be permitted to control the situation provided they have evidence that the birds are in fact carrying parasitic disease organisms. Anglers on the other hand – please be prepared to spare rare species.

Kingfishers, herons and otters, dear to lovers of wild-life, are sometimes frowned on by fishermen because they are regarded as fish thieves. Indeed they do take fish but, if present only in small numbers, they are not likely to do serious damage to a fishery; rather they might well do good by picking off the less speedy and healthy of the fishes and perhaps helping to control numbers where the problem is an overstocked water.

As far as the general activities of anglers are concerned they are surely a quiet and peace-loving fraternity – or at least they should be if they want to catch anything! Perhaps one danger to wild-life which could be avoided is the discarding of nylon line. If lengths of this are left lying on the ground, or in bushes, birds can so easily become entangled. A plea then always to take your unwanted line home with you. Another thing to take home of course is litter! Anglers are rather inclined to leave bread wrappings around and this does tend to spoil the pleasures of other country-lovers.

There seems to me no reason why aquatic ecosystems cannot be managed for the benefit of anglers, nature lovers and bird-watchers alike. All that is required is a will to approach things in a scientific way so that our knowledge of natural systems can be increased, and a will to co-operate in their management with all those who have an interest in the countryside. More liaison between angling clubs and such organisations as the county trusts for nature conservation would be a good thing here. Indeed, as all country lovers and out-door sportsmen are being forced into a backs-to-the-wall position in the face of expanding urbanisation and advanced mechanised agriculture we must work together to conserve the delights of the countryside which so many of us hope to continue to enjoy.

Books recommended for further reading

For books on classification and identification of plants and animals see end of Chapter 1.

Moderately non-technical books for general reading:
J Clegg, 1973. The Freshwater Life of the British Isles. Warne.
W E Frost & M E Brown, 1967. The Trout. New Naturalist, Fontana.
J R Harris. An Angler's Entomology. New Naturalist, Collins.
J Jones, 1959. The Salmon. New Naturalist, Collins.
T T Macan, 1973. Ponds and Lakes. George Allen & Unwin.
T T Macan & E D Worthington, 1972. Life in Lakes and Rivers. New Naturalist, Fontana.
K Mellanby, 1969. Pesticides and Pollution. New Naturalist, Fontana.
D H Mills, 1972. An Introduction to Freshwater Ecology. Oliver & Boyd.
E J Popham, 1955. Some Aspects of Life in Freshwater. Heinemann.
T O Robson, 1973. MAFF Bulletin 194. The Control of Aquatic Weeds. HMSO.
M E Varley, 1967. British Freshwater Fishes. Fishing News (Books) Ltd.

Rather more technical books for further study and reference:
G W Bennett, 1971. Management of Lakes and Ponds. Van Nostrand Reinhold Co.
V A Dogiel, G K Petrushevsky & Yu I Polyansky, 1961. Parasitology of Fishes. Oliver & Boyd.
S M Evans, 1970. The Behaviour of Birds, Mammals and Fish. Heinemann.
F R Harden Jones. Fish Migration. Arnold.
Hoar & Randall. Fish Physiology (6 volumes). Academic Press.
M Huet. Textbook of Fish Culture, Breeding and Cultivation of Fish. Fishing News (Books) Ltd.
H B N Hynes, 1966. The Biology of Polluted Waters. Liverpool University Press.

H B N Hynes, 1971. The Ecology of Running Waters. Liverpool University Press.

E J Kormondy, 1969. Concepts of Ecology. Prentice-Hall.

N B Marshall, 1965. The Life of Fishes. Weidenfeld & Nicholson.

N B Marshall, 1971. Explorations in the Life of Fishes. Harvard Books in Biology No. 7. (Cambridge Massachusetts.)

R C Muirhead Thomson, 1971. Pesticides and Freshwater Fauna. Academic Press.

W E Ricker, 1968. Methods for Assessment of Fish Production in Fresh Waters. IBP Handbook No. 3, Blackwell Scientific Publications.

L Mawdesley Thomas, 1972. Diseases of Fish. Zool. Society Symposium No. 30, Academic Press.

A H Weatherly, 1971. The Growth and Ecology of Fish Populations. Academic Press.

Index

Abstraction of water, 114—115
Acanthocephalan, 101
Acclimatisation, 55—56
Age determination of fish, 82—85, 90
Algae, 8, 25, 36
  see also Blanket weed
  and eutrophication, 106
  in food webs, 40—46
  production of, 88—89
  in still and flowing waters, 36—37
  and weed problems, 95—96
Allen Curve, 89
Amino acid, 2, 3
Ammonia, 107
Ampulla, 69, 70
Anchor worm, 101
Animal(s), 8, 9
  in food webs, 39—42
  see also Invertebrates, Vertebrates
  productivity, 88—89
  in still and flowing waters, 36—37
Annelida, 9, 111
Antibiotic, 103
Antibodies, 102
Anus, 13, 16, 17
Aorta, 14
Appetite, 20
Argulus, 101
Arsenic, 107
Artery, 14
Asellus, 108, 110—112
Atom, 2, 3

Bacteria, 9, 39
  causing decay and de-oxygenation, 27, 41, 42, 106, 107
  causing disease, 98, 102—104
Baetis, 110, 111
Bait, 47—49, 71, 77—79
Balance, 70, 73
Barbel, 37, 61
Basal metabolic rate, 34, 54, 55, 57, 58

Beetle larvae, 42, 111
Behaviour, 63—81
  inherited, 74—75
  learned, 75—76
Biological control of disease, 104
Biotic Index, 112, 113
Biotope, 10
Bird(s), 99, 100, 104, 116, 117
Black fly larva, 111
Black spot, 101
Blanket weed, 8, 41, 95
  see also *Algae*
Blood, 14—17, 66
Blood vessel, 14, 15, 66
Bloodworm, 112, 113
Bluestone, 95
BOD (Biochemical oxygen demand), 106
Bone, ear, 70, 83
  opercular, 83, 84
Brain, 13, 64, 65, 70
Bream, 36, 37, 46, 48, 55, 57, 58, 60, 61, 74, 82, 84, 114
  bluegill, 91, 92
Breeding habits, 61—62
Bullhead, 60

Caddis larvae, 45, 47, 110—112
Cadmium, 107
Calcium, 29, 31, 70
Canadian pondweed, 40, 95
Canal, 34, 38
  lateral line, 68
  semi-circular, 69, 70, 73
Capillaries, 15
Carbohydrate, 7, 16
Carbon, 2, 3, 29, 88
Carbonate, 29, 31
Carbon dioxide, 3, 4, 28, 39, 40
Carnivore, 39—42, 44, 46, 48—51
Carp, 36, 46, 53—55, 57—61, 70, 77, 78, 80, 102, 103
  grass, 95—96
Catfish, 80

Cell, 5, 14, 19, 55, 65, 68, 107
  sensory, 64, 68–71, 74
Cellulose, 5
Char, 36, 54
Chara, 45
Check, annual, 83, 84
  double, 85
  false, 85
Chemical therapy of disease, 103
Chironomidae, 111
Chironomid larvae, 110–112
Chloride, ferric, 29
Chlorophyll, 4, 88
Chromium, 107
Chub, 37, 54
Cilia, 71, 101
Class, 8–10
Classification, 11, 21
Closed season, 79
Coarse fish, 18, 36, 37, 44, 47, 48, 52, 57, 83, 93, 100, 108, 113
Cold blooded, 34, 54
Coleoptera, 111
Columnaris, 102
Communication, of fish, 79–81
Community, 11
Competition, 91
Conditioning, 73–77
Conductivity, of water, 29
Conjunctiva, 66
Conservation, 115–119
Copper, 107
Copper sulphate, 95
  see also Bluestone
Cornea, 66
Cotton wool disease, 102
Courtship, 75
Crustacea, 9
  see also Crustacean
Crustacean, 9, 41, 46, 61, 101, 111
Cupula, 68–70
Curly leaved pondweed, 95
Current, 71, 74, 93
Cyanide, 107
Cyprinid, 13, 16, 44, 54, 55, 61
Cyst, 101
Cytoplasm, 5–7

Dace, 37, 46, 60, 61
Decay, 27, 40–42, 106, 116
Decomposition, see Decay
Decomposers, 49
De-Oxygenation, 26, 27, 116
  in hypolimnion, 34
  see also Oxygen
  in polluted water, 106, 112
Depth, 31–34, 36

Detergent, 107
Detritus, 40–42
Diatomeae, 45
  see also Diatoms
Diatoms, 41, 45, 46, 61
Digestion, 16
Digestive system, 16
Disease, 97–105, 117
Diversity Index, 113
DNA (Deoxyribonucleic acid), 6
Dragonfly larvae, 41, 112
Drift fauna, 42

Ear, 64, 69–71
Ecosystem, 11, 35, 49, 50, 117
Eel, 25, 46, 60
Efficiency, of energy transfers, 50–51
Effluent, 57, 106, 108, 112
Eggs, of fish, 18, 61, 75, 89
  of parasite, 99, 100
Electrical conductivity of water, 29
Electrical field, production of and sensitivity to, 65, 80, 81
Electro-fishing, 86
Energy, 2, 3, 6, 7, 14, 39, 89
  flow, 49–51
Enzyme, 16
Ephemeridae, 66
Ephemeroptera, 110–112
Epilimnion, 32, 33
Eristalis (rat-tailed maggot), 110
Euryhaline, 25
Eutrophic, 52, 106
Eutrophication, 106–107
Evolution, 7, 8, 9
Excretion, 17
Experiment, 22–24, 26, 73–74, 77, 91–92
Eye, 64–68, 97–100
Eyeball, 65
Eyelids, lack of in fish, 66, 68

Falciform process, 66
Family, 10
Fat, 5, 7, 16
Feeding habits, 44–46, 57–58
  see also Food
Fertilisation, in reproduction, 18
Fertilisers, 31, 92, 93–95, 106
  see also Nitrates and Phosphates
Fin rot, 102
Fins, 12, 17
Flagellates, 41
Flatworms, 44, 111
Fluke, 8, 9, 98, 101–104, 106
Fluke, eye, 98–101

Fly, in fly fishing, 66, 68, 78
Focusing, 65, 66, 99
Food, 2, 4–6, 16, 19, 39–52, 55, 89, 90, 95, 116
  in behaviour experiments, 73–77
  see also Feeding habits
Foodweb, 39–42, 46, 50
Fry, 18, 48, 61, 73–74, 89–93, 108
Fuel, 1, 2, 4, 7, 39, 51
Fungi, 8, 27, 41, 42
Furunculosis, 102

Game fish, 44, 52, 83, 108
Gammaridae, 45
Gammarus, 110–112
Genus, 10, 11
Gill cover, 12, 14, 15, 26, 83
  see also Operculum
Gills, 14, 15, 24, 26, 97, 101, 107
Glands, 18, 19
Glucose, 2, 3, 4, 14
Goldfish, 10, 24, 26, 54, 55, 56, 58, 60, 61, 73, 74, 76–79
Grayling, 37, 54, 55
Growth, 19–20, 39, 82–85, 90–94, 108
Gudgeon, 46, 55
Gut analysis, 42–44

Habitat, 11
Haemoglobin, 15
Handling of fish, 72–73, 104
Hatcheries, 50–51, 93, 103
Hearing, 69–71
Heart, 12, 13
Hemiptera, 111
Herbicides, 95, 107
Herbivore, 39–42, 44, 46, 49, 50
Hirudineae, 111,
Hog louse, 42, 111
  see also Water slater
Hook(s), 16, 73, 77, 78, 104
Hormones, 18, 19
Host, of parasite, 98–104
Hydracarina, 111
Hydrochloric acid, 16
Hydrogen, 2, 3, 29
Hypolimnion, 32, 33

Ice, 28, 31
Ichthyophthiriasis, 101
Ichthyophthirius, 101
IDC (Infectious dropsy of carp), 102
Identification, 11, 21
Index, Biotic, 112, 113
  Diversity, 113
  Lincoln, 86

Indicators of pollution, 112, 113
Insects, 9, 41, 46
Instinct, 75, 79
Intestine, 12, 13, 16, 42
Invertebrates (animals without backbones), 36, 37, 47, 95, 98, 103, 108
Iris, 66
Iron, 3, 29

Keep net, 27, 104
Key, for identification, 12
Kidney(s), 13, 17, 24

Lake(s), 30–36, 57, 58, 61, 93
Larva(e), alder fly, 111
  beetle, 42, 111
  black fly, 111
  caddis, 111, 112
  dragonfly, 41, 42
  midge, 111
  of parasite, 99, 100
Lateral line, 12, 64, 65, 68–70
Lead, 107
Learning, 75–79
Leaves, 41–42
Leech, 41, 44, 111
Length of fish, 85
Lens, 65, 66
Lernaea, 101
Ligament, 65, 66
Light rays, 65, 67
  waves, 64
Ligula, 101
Limestone, 31, 37, 93–94
Liming, 94
Lincoln Index, 86
Line, 68, 117
Litter, 117
Liver, 13
Loach, 46, 60
Louse (lice), 98, 101
  see also Hog louse
Lury, de, 86

Magnesium, 5, 29
Maintenance requirements, 20
Management of fisheries, 46–47, 82, 89–96, 115–117
Marking, 87
Mark-recapture method of population estimation, 86–87
Mating of stickleback, 80
Mayfly nymphs, 109–113
Memory, 76
Microbes, 102
  see also Bacteria and Fungi

Micro-organisms, 102
    see also Bacteria and Fungi
Midge larvae, 44, 111
Migration, 72
Milt, 18
Minerals, 42, 49
    see also Salts, Nutrient salts
Mineral salts, 42, 49
    see also Salts, Nutrient salts
Minnow, 37, 60, 80
Molecule, 2, 3
Mollusca, 9, 45
Molluscs, 45, 46, 90, 111
    see also *Mollusca* and Snails
Mortality, 89–90, 92
Mouth, 13, 26, 71
Mouth fungus, 102
Movement, 17, 58–61
Muscle, 14, 17, 65, 66
Mycobacterium, 102

Nasal sac, 12, 64, 71
Natural selection, 7
Nematode worm, 101
Nerve(s), 64
    fibres, 66, 68–72, 74
    optic, 65, 66
Nervous system, 19, 64, 65, 72–74
Nest of stickleback, 75
Netting, 86
Neuroptera, 111
Nickel, 107
Nitrate(s), 5, 29
    in fertilisers, 94, 106
Nitrogen, 2, 3, 29, 39, 94
Nucleic acid, 6, 7
    see also DNA
Nucleus, 5–7
Numbers in population, 82
    methods of estimating, 86–88
    in relation to stocking, 91–93
Nutrient salts, see Salts, Minerals,
    Mineral salts
Nymph(s), 66, 68, 73
    mayfly, 109–113
    stonefly, 109–112

Oesophagus, 13, 80
Olfactory organs, 71–72
    see also Smell and Taste
Oligotrophic, 52
Omnivore, 40, 46, 48
Operculum, 12, 14, 15, 26, 83
    see also Gill cover
Order, 10, 11
Organ, 12, 13
Ossicle, 70

Otolith, 69, 70, 83
    see also Bone, ear
Ovary, 13, 18
Oxygen, 1, 2, 29, 66
    demand, 106
    see also De-oxygenation
    dissolved in water, 25–29, 36, 37
    in hypolimnion, 34
    lack of in polluted water, 112,
        114
    and metabolism, 55, 58
    in photosynthesis, 4
    in respiration, 3, 7, 14, 15

Pain, 72–73
Parasite, 98–104, 116, 117
Perch, 36, 37, 44, 45, 46, 48, 54, 55,
    57, 58, 60, 82, 100
Pesticides, 107
pH, 29–31, 36, 37, 95
Pharynx, 13
Phenols, 107
Phosphate(s), 5
    in fertilisers, 94, 106, 107
    see also Salts, Minerals
Phosphorus, 5, 29, 39, 94
Photosynthesis, 4, 5, 7, 27, 28, 35,
    41, 39, 88
Phylum, 8–10
Phytoplankton, 41, 51, 91
Pike, 36, 37, 44, 46, 48, 55, 57, 60,
    78, 93, 100
Pisces, 8, 11
Plankton, 36, 41, 42, 88
Plants, and conservation, 116
    control of weed plants, 95, 96
    dead plants and pollution, 106
    in food webs, 39–42, 47, 50
    kinds of, 8, 9
    and the manufacture of food,
        4–6
    and oxygen, 27, 28
    productivity of, 88–90
    in still and flowing waters, 36–37
Platyhelminthes, 9, 111
Plecoptera, 110–112
Poison, 107, 114
Pollution, 105–114
    organic, 106–107
    thermal, 107–108
    toxic chemical, 107
Pond, 34, 38, 91–96
Pondweed, canadian, 40, 95
    curly leaved, 95
    floating leaved, 40
Pool, 31, 34, 42, 53, 61, 90–96, 112
Population, 11

124

Population numbers, see Numbers
Potassium, 5, 29, 94
Pox, of carp, 102
Predation, 93
Pressure, 64, 69, 71
   barometric, 72
Producers, 39, 40, 49
Productivity, 52, 88–95, 106
Protein, 2, 7, 16, 39
Protozoa, 9, 41, 98, 101, 106
Pupil, 65, 66

Redd, 75
Reeds, 40, 89
Reflex, 73–75
Regional Water Authority, 104, 112, 113
Reproduction, 18
   see also Breeding habits
Reservoir, 36, 38
Respiration, 3, 7, 14, 27, 49
Response, 73–76, 79
Retina, 65, 66, 74
River(s), 34, 37, 38, 42, 57, 61, 66, 71, 103, 114, 115
Roach, 20, 36, 37, 44, 45, 46, 55, 56, 57, 60, 82, 90, 114
Roe, 18
Rotifers, 9, 41
Roundworms, 101
Rudd, 37, 48, 55, 57, 60, 82
Ruffe, 55

Sacculus, 69, 70
Salinity, 64
Salmon, 18, 25, 54, 55, 60, 103, 105, 108
Salmonid(s), 36, 37, 44, 52, 55, 58, 59, 60, 61, 75, 93, 100, 108
Salts, nutrient, 5, 28, 29, 31, 39, 42, 49, 52, 91, 93, 94, 106
   see also Minerals, Mineral salts
Sampling, 86, 87, 94
Scales, 24, 83, 85
Scale reading, 83–85
Sea horse, 80
Seiche, 33
Senses, 64–74
Sense organs, 64–74
Sewage, 105–107
Shapes of fish, 58–61
Shrimp(s), 25, 41, 42, 45, 111, 112
Signal, 79
Silage, 106
Silicon, 29
Simulidae, 111
Size of fish, 91–92

Skeleton, 17
Skin, 24, 68, 71, 100–104
Skull, 69
Slime, 97
Smell, 71, 72, 77, 79
   see also Olfactory organs, Nasal sacs
Snails, 9, 41, 42, 100, 104, 111
   see also Molluscs
Sodium, 29
Sound, 64, 69, 70, 80
   production, 80
Spawn, 18
Species, 10, 11
Sperm, 18
Spinal cord, 13, 64, 65
Spinner, 48, 78
Spores, 101
Starch, 16
Stenohaline, 24
Stickleback, 18, 25, 46, 75
Stimulus, 73, 75, 76
Stocking, 91–93
Stomach, 13, 16, 42
Stonefly nymphs, 109–112
Stratification of lakes, 32, 33, 36, 37
Stream(s), 31, 34, 37, 38, 42, 53, 57, 58, 61, 62, 71, 108, 114, 115
Streamlining, 58, 59
Stunting, 20, 44, 82, 90–91
Sub-class, 11
Sub-species, 10
Sugar, 5, 16, 40
Sulphate, 29
Sunlight, 4, 5, 7, 27, 28, 35, 39, 49, 95
   see also Light waves and rays
Swim-bladder, 13, 58, 70, 80
Swimming, 17, 58–61

Tackle, 77, 79, 104
Tail, 17, 59, 60, 61
   rot, 102
Tank experiments, 73–77
Tapeworm, 8, 9, 101
Taste, 71–72
   see also Olfactory organs
Teleost, 11–13, 66, 80
Temperature, and de-oxygenation, 25–27
   detection of changes, 64, 72
   and disease, 103
   and growth, 20
   lethal, 55–57
   and metabolism, 54–58
   and stratification, 31–35

and thermal pollution, 108
toleration of extremes, 93
and toxic pollution, 114
Tench, 36, 37, 46, 54, 55, 57, 58, 59
Testis, 13, 18
Thermocline, 32, 33
Toxicity, 107, 113, 114
Training, 75, 76
Trees, round edges of angling waters, 41, 116
Trichoptera, 45, 110, 111
Trigger fish, 80
Trout, 18, 19, 20, 36, 37, 44, 46, 47, 48, 53, 54, 55, 58, 60, 61, 66, 68, 71, 74, 78, 93, 94, 102, 103, 114
brown, 19, 20
rainbow, 54, 76, 113
sea, 25
speckled, 54, 56
Tuberculosis, 102
Tubificid worms, 110–112

UDN (Ulcerative dermal necrosis), 103
Ulcer, 97, 103
Urinary bladder, 13, 17
Urine, 17
Urinogenital opening, 13, 17
Utriculus, 69

Vein, 14
Vertebra, 13
Vertebrates (animals with backbones), 69
Vibration, 69
Viruses, 9, 98, 102, 104
Vision, 65–68
colour, 68, 74

Water, abstraction of, 114–115
density of, 25
formula of, 3
hard and soft, 28, 52
pollution of, 105–114
in respiration and photosynthesis, 3, 4
Water lily, 40, 89
mites, 111
slater, 41, 42, 108, 111, 112
see also Hog louse
starwort, 40
Weed, 95–96
see also Pondweed and Plants
Weight, 85
Whitefish, 36
White Grunt, 80
White spot, 101
Worms, 9, 41, 44, 98–101, 110–112

Zinc, 107
Zooplankton, 41, 51

List of other books published by Fishing News (Books) Limited

Free catalogue available on request

A living from lobsters
British freshwater fishes
Coastal aquaculture in the Indo-Pacific region
Commercial fishing methods
Control of fish quality
Culture of bivalve molluscs
Eel capture, culture, processing and marketing
Eel culture
Escape to sea
European inland water fish: a multilingual catalogue
FAO catalogue of fishing gear designs
FAO investigates ferro-cement fishing craft
Farming the edge of the sea
Fish and shellfish farming in coastal waters
Fish catching methods of the world
Fish farming international 1, 2 and 3
Fish inspection and quality control
Fisheries oceanography
Fishery products
Fishing boats of the world 1
Fishing boats of the world 2
Fishing boats of the world 3
Fishing ports and markets
Fishing with electricity
Freezing and irradiation of fish
Handbook of trout and salmon diseases
Handy medical guide for seafarers
How to make and set nets
Inshore craft of Britain in the days of sail and oar
Inshore fishing: its skills, risks, rewards
International regulation of marine fisheries: a study of regional
    fisheries organizations

Introduction to trawling
Japan's world success in fishing
Marine pollution and sea life
Mechanization of small fishing craft
Mending of fishing nets
Modern deep sea trawling gear
Modern fishing gear of the world 1
Modern fishing gear of the world 2
Modern fishing gear of the world 3
Modern inshore fishing gear
More Scottish fishing craft and their work
Multilingual dictionary of fish and fish products
Netting materials for fishing gear
Power transmission and automation for ships and submersibles
Refrigeration on fishing vessels
Seafood fishing for amateur and professional
Ships' gear 66
Sonar in fisheries: a forward look
Stability and trim of fishing vessels
Testing the freshness of frozen fish
Textbook of fish culture; breeding and cultivation of fish
The fertile sea
The fish resources of the ocean
The fishing cadet's handbook
The lemon sole
The marketing of shellfish
The seine net: its origin, evolution and use
The stern trawler
The stocks of whales
Trawlermen's handbook
Tuna: distribution and migration
Underwater observation using sonar